T0269281

SpringerBriefs in Electrical and Computer Engineering

More information about this series at http://www.springer.com/series/10059

Awais Khawar · Ahmed Abdelhadi
T. Charles Clancy

MIMO Radar Waveform Design for Spectrum Sharing with Cellular Systems

A MATLAB Based Approach

Awais Khawar
Virginia Tech
Arlington, VA
USA

T. Charles Clancy
Virginia Tech
Arlington, VA
USA

Ahmed Abdelhadi
Virginia Tech
Arlington, VA
USA

ISSN 2191-8112 ISSN 2191-8120 (electronic)
SpringerBriefs in Electrical and Computer Engineering
ISBN 978-3-319-29723-1 ISBN 978-3-319-29725-5 (eBook)
DOI 10.1007/978-3-319-29725-5

Library of Congress Control Number: 2016930679

Printed on acid-free paper

This Springer imprint is published by SpringerNature
The registered company is Springer International Publishing AG Switzerland

Preface

Recent years have seen tremendous growth in use of radio frequency spectrum especially by commercial cellular operators. Ubiquitous use of smartphones and tablets is one of the reasons behind an all-time high utilization of spectrum. As a result, cellular operators are experiencing a shortage of radio spectrum to meet bandwidth demands of users. On the other hand, spectrum measurements have shown that much spectrum not held by cellular operators is underutilized even in dense urban areas. This has motivated shared access to spectrum by secondary systems with no or minimal impact to incumbent systems. Spectrum sharing is a promising approach to solve the problem of spectrum congestion as it allows cellular operators access to more spectrum in order to satisfy the ever-growing bandwidth demands of commercial users. The US spectrum regulatory bodies, the Federal Communications Commission (FCC) and the National Telecommunications and Information Administration (NTIA), are working on an initiative to share 150 MHz of spectrum, held by federal agencies, in the band 3550–3700 MHz with commercial wireless operators. This band is primarily used by the Department of Defense for air, ground, and shipborne radar systems that are critical to national defense.

Radars operating in this band are a major source of interference to communication systems. However, radar waveform can be transformed in such a way that it does not interfere with communication systems. This is accomplished by projecting the radar signal onto the null space of the wireless channel between radar and communication system. This book discusses two different types of radar waveforms that are designed specifically for congested RF spectrum environments, thus, enabling simultaneous operation of radar and communication systems.

Arlington, VA, USA
November 2015

Awais Khawar
Ahmed Abdelhadi
T. Charles Clancy

Acknowledgments

The work reported in this book was supported by DARPA under the SSPARC program. Contract Award Number: HR0011-14-C-0027. We are grateful to our sponsors for funding spectrum sharing research between radar and communication systems.

Contents

List of Figures

List of Algorithms

Chapter 1
Introduction

An interesting concept for next generation of radars is multiple-input multiple-output (MIMO) radar systems; this has been an active area of research for the last couple of years [1]. MIMO radars have been classified into widely-spaced [2], where antenna elements are placed widely apart, and colocated [3], where antenna elements are placed next to each other. MIMO radars can transmit multiple signals, via their antenna elements, that can be different from each other, thus, resulting in waveform diversity. This gives MIMO radars an advantage over traditional phased-array radar systems which can only transmit scaled versions of single waveform and, thus, can not exploit waveform diversity.

Waveforms with constant-envelope (CE) are very desirable, in radar and communication system, from an implementation perspective, i.e., they allow power amplifiers to operate at or near saturation levels. CE waveforms are also popular due to their ability to be used with power efficient class C and class E power amplifiers and also with linear power amplifiers with no average power back-off into power amplifier. As a result, various researchers have proposed CE waveforms for communication systems; for example, CE multi-carrier modulation waveforms [4], such as CE orthogonal frequency division multiplexing (CE-OFDM) waveforms [5]; and radar systems, for example, CE waveforms [6], CE binary-phase shift keying (CE-BPSK) waveforms [7], and CE quadrature-phase shift keying (CE-QPSK) waveforms [8].

Existing radar systems, depending upon their type and use, can be deployed anywhere between 3 and 100 GHz of radio frequency (RF) spectrum. In this range, many of the bands are very desirable for international mobile telecommunication (IMT) purposes. For example, portions of the 700–3600 MHz band are in use by various second generation (2G), third generation (3G), and fourth generation (4G) cellular standards throughout the world. It is expected that mobile traffic volume will continue to increase as more and more devices will be connected to wireless networks. The current allocation of spectrum to wireless services is inadequate to support growth in traffic volume. A solution to this spectrum congestion problem was

A. Khawar et al., *MIMO Radar Waveform Design for Spectrum Sharing with Cellular Systems*, SpringerBriefs in Electrical and Computer Engineering,
DOI 10.1007/978-3-319-29725-5_1

presented in a report by President's Council of Advisers on Science and Technology (PCAST), which advocated to *share* 1000 MHz of government-held spectrum [9]. As a result, in the United States (U.S.), regulatory efforts are underway, by the Federal Communications Commission (FCC) along with the National Telecommunications and Information Administration (NTIA), to share government-held spectrum with commercial entities in the frequency band 3550–3650 MHz [10]. In the U.S., this frequency band is currently occupied by various services including radio navigation services by radars. The future of spectrum sharing in this band depends on novel interference mitigation methods to protect radars and commercial cellular systems from each others' interference [11–15]. Radar waveform design with interference mitigation properties is one way to address this problem, and this is the subject of this book.

1.1 Spectrum Sharing Efforts Between Radar and Communication Systems

A study by the NTIA evaluated sharing of radar band with WiMAX systems and found that in order to protect WiMAX systems from radar interference huge exclusion zones upto tens of kilometers are required [16]. This is due to high signal power used by radars and high-peak sidelobes which saturate communication system receivers, which are traditionally designed to handle power levels in watts rather than kilo watts or mega watts. Such high peak powers are typical of airport surveillance radars, weather radars, and military phased array radars such as SPY-1 radar of Aegis system. On the other hand, due to highly sensitive radar receivers, designed to detect even the faintest of returned signal, has in the past mandated for exclusive rights to radio spectrum allocations since its operation can be affected by commercial wireless system interference [16, 17].

The heterogeneous nature of devices sharing an RF band, radar and cellular system in our case, dictates the need for electromagnetic interference (EMI) mitigation tactics for both systems since traditional interference mitigation tactics are meant for exclusive use of the same RF band. The emission pattern, both in space and time, of radar is significantly different from communication system. This point is also validated from a study by the NTIA, showing that radar receivers handle noise like interference from communications systems differently than the interference from other radars with former having detrimental effect on radar due to its continuous wide-band nature than the low duty cycle radar waveforms [16].

In the past, it has been made possible for wireless systems to share government bands such that they operate under a low-power constraint in order to protect incumbents from interference. Example includes: Wi-Fi and Bluetooth in the 2450–2490 MHz band, wireless local area network (WLAN) in the 5.25–5.35 and 5.47–5.725 GHz radar bands [18], and recently the FCC has proposed small cells, i.e. wireless base stations operating on a low power, to operate in the 3550–3650 MHz radar band [10].

The 3550–3650 MHz band, currently used for military and satellite operations, is a possible candidate for spectrum sharing between military radars and broadband wireless access (BWA) communication systems such as LTE and WiMAX, according to the NTIA's 2010 Fast Track Report [19]. Electromagnetic interference to military radar operations is expected from spectrum sharing. However, one simply can't relocate these federal radar systems to other bands since the nature of the said band contains many frequencies which work best for highly sensitive fixed, airborne, and maritime radar systems and are essential for superior performance. Moreover, cost to relocate can be unbearable. The problem of EMI mitigation is possible due to advancements in transmitter and receiver design technology, of cellular systems, which has made real-time spectrum reassignment possible.

In spectrum sharing perspective between radars and communication systems, EMI needs to be mitigated at both the systems. Communication systems due to their advancements give more freedom to mitigate interference from radar systems. For example, in order to counter radar interference on WiMAX systems, interference mitigation in four domains namely space, time, frequency, and system-level modification is proposed by Lackpour et al. [20]. Radar systems due to their sensitivity are more susceptible to interference from communication systems. So far, as previously discussed, in order to protect radar operations, communication systems operate on a low-power basis to avoid interference to radars or operate by sensing the availability of radar channel at a power level which doesn't exceed the allowed interference limit [18, 21].

Radar systems are also evolving and with recent trend in design of MIMO radars and cognitive radars, radar systems are becoming more resilient in handling interference and jamming as they are more aware about their radio environment map (REM). This has motivated researchers to propose beamforming approaches to mitigate interference from wireless communication systems to MIMO radar [22]. In addition, spatial domain can also be used to mitigate MIMO radar interference to wireless communication system. One such technique was proposed by Shabnam et al. [23] which projected radar signal onto the null space of interference channel between MIMO radar and MIMO communication system. Moreover, radar waveforms can also designed such that they don't cause interference to cellular systems, in addition to meeting their mission objectives [24].

1.2 Waveform Design for Congested Spectrum

Transmit beampattern design problem, to realize a given covariance matrix subject to various constraints, for MIMO radars is an active area of research; many researchers have proposed algorithms to solve this beampattern matching problem. Fuhrmann et al. proposed waveforms with arbitrary cross-correlation matrix by solving beampattern optimization problem, under the constant-modulus constraint, using various approaches [25]. Aittomaki et al. proposed to solve beampattern optimization problem under the total power constraint as a least squares problem [26]. Gong et al.

proposed an optimal algorithm for omnidirectional beampattern design problem with the constraint to have sidelobes smaller than some predetermined threshold values [27]. Hua et al. proposed transmit beampatterns with constraints on ripples, within the energy focusing section, and the transition bandwidth [28]. However, many of the above approaches don't consider designing waveforms with finite alphabet and constant-envelope property, which is very desirable from an implementation perspective. Ahmed et al. proposed a method to synthesize covariance matrix of BPSK waveforms with finite alphabet and constant-envelope property [7]. They also proposed a similar solution for QPSK waveforms but it did not satisfy the constant-envelope property. Similarly, Jardak et al. proposed to generate infinite alphabet constant envelope waveform [29]. A method to synthesize covariance matrix of QPSK waveforms with finite alphabet and constant-envelope property was proposed by Sodagari et al. [8]. However, they did not prove that such a method is possible. It was shown that it is possible to synthesize covariance matrix of QPSK waveforms with finite alphabet and constant-envelope property [30].

As introduced earlier due to the congestion of frequency bands future communication systems will be deployed in radar bands. Thus, radars and communication systems are expected to share spectrum without causing interference to each other. For this purpose, radar waveforms should be designed in such a way that they not only mitigate interference to them but also mitigate interference by them to other systems [23, 31]. Transmit beampattern design by considering the spectrum sharing constraints is a fairly new problem. Sodagari et al. have proposed BPSK and QPSK transmit beampatterns by considering the constraint that the designed waveforms do not cause interference to a single communication system [8]. This approach was extended to multiple communication systems, cellular system with multiple base stations, by Khawar et al. for BPSK transmit beampatterns [24, 32].

1.3 System Models

In the following sections, we describe radar system model, cellular system model, interference channel model, and spectrum sharing environment between radar and communication systems.

1.3.1 Radar Model

We consider waveform design for colocated MIMO radar mounted on ship. The radar has M colocated transmit and receive antennas. The inter-element spacing between antenna elements is on the order of half the wavelength. The radars with colocated elements give better spatial resolution and target parameter estimation as compared to radars with widely spaced antenna elements [2, 3]. A detailed discussion on radar signal type will be provided in Chap. 2 for BPSK waveforms and in Chap. 3 for QPSK waveforms.

1.3.2 Cellular System Model

We consider MIMO cellular system, with $\mathcal{K}$ base stations, each equipped with N_{BS} transmit and receive antennas, with the ith BS supporting $\mathcal{L}_i$ user equipments (UEs). Moreover, the UEs are also multi-antenna systems with N_{UE} transmit and receive antennas. If $\mathbf{s}_j(n)$ is the signals transmitted by the jth UE in the ith cell, then the received signal at the ith BS receiver can be written as

$$\mathbf{y}_i(n) = \sum_j \mathbf{H}_{i,j}\,\mathbf{s}_j(n) + \mathbf{w}(n), \qquad \text{for } 1 \leqslant i \leqslant \mathcal{K} \text{ and } 1 \leqslant j \leqslant \mathcal{L}_i$$

where $\mathbf{H}_{i,j}$ is the channel matrix between the ith BS and the jth user and $\mathbf{w}(n)$ is the additive white Gaussian noise.

1.3.3 Interference Channel Model

In our spectrum sharing model, radar shares $\mathcal{K}$ interference channels with cellular system. Let's define the ith interference channel as

$$\mathbf{H}_i \triangleq \begin{bmatrix} h_i^{(1,1)} & \cdots & h_i^{(1,M)} \\ \vdots & \ddots & \vdots \\ h_i^{(N^{\mathrm{BS}},1)} & \cdots & h_i^{(N^{\mathrm{BS}},M)} \end{bmatrix} \quad (N_{\mathrm{BS}} \times M) \tag{1.1}$$

where $i = 1, 2, \ldots, \mathcal{K}$, and $h_i^{(l,k)}$ denotes the channel coefficient from the ith antenna element at the MIMO radar to the lth antenna element at the ith BS. We assume that elements of $\mathbf{H}_i$ are independent, identically distributed (i.i.d.) and circularly symmetric complex Gaussian random variables with zero-mean and unit-variance, thus, having i.i.d. Rayleigh distribution. In addition to this interference channel, other types of interference channels can be considered for spectrum sharing scenarios. A detailed discussion on various interference channels can be found in [33–35].

1.3.4 Cooperative RF Environment

Spectrum sharing between radars and communication systems can be envisioned in two types of RF environments, i.e., military radars sharing spectrum with military communication systems, we characterize it as *Mil2Mil* sharing and military radars sharing spectrum with commercial communication systems, we characterize it as *Mil2Com* sharing. In *Mil2Mil* or *Mil2Com* sharing, interference-channel state information (ICSI) can be provided to radars via feedback by military/commercial communication systems, if both systems are in a frequency division duplex (FDD) configuration [36]. If both systems are in a time division duplex configuration, ICSI can be obtained via exploiting channel reciprocity [36]. Regardless of the configuration of radars and communication systems, there is the incentive of zero interference, from radars, for communication systems if they collaborate in providing ICSI. Thus, we can safely assume the availability of ICSI for the sake of mitigating radar interference at communication systems.

1.3.5 Spectrum Sharing Scenario

Considering the coexistence scenario in Fig. 1.1, where the radar is sharing $\mathcal{K}$ interference channels with the cellular system, the received signal at the ith BS can be written as

$$\mathbf{y}_i(n) = \mathbf{H}_i \mathbf{x}(n) + \sum_j \mathbf{H}_{i,j}\, \mathbf{s}_j(n) + \mathbf{w}(n) \tag{1.2}$$

In order to avoid interference to the ith BS, the radar shapes its waveform $\mathbf{x}(n)$ such that it is in the null-space of $\mathbf{H}_i$, i.e. $\mathbf{H}_i\mathbf{x}(n) = \mathbf{0}$.

1.4 Spectrum Sharing via a Projection Based Scheme

In wireless communications, multiple access schemes have played a vital role in enabling multiple users access spectrum simultaneously. Some of the very common and widely used multiple access schemes include frequency division duplexing

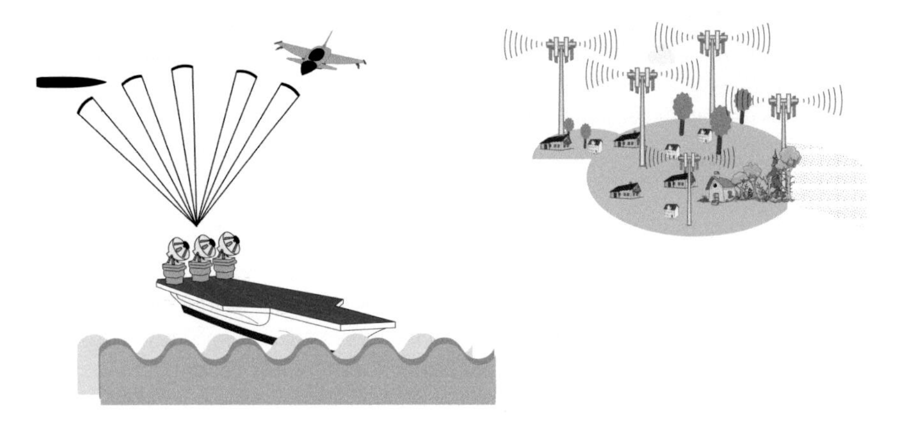

Fig. 1.1 Spectrum sharing scenario: Seaborne MIMO radar detecting a point target while simultaneously sharing spectrum with a MIMO cellular system

(FDD) and time division duplexing (TDD). However, the emergence of MIMO systems made it possible to exploit spatial domain. This gave rise to spatial division duplexing (SDD) in which users coexist in the same channel by exploiting orthogonal spatial dimensions, that are not in use by the other user. This results in an interference free environment and is also known as null space based coexistence method.

Projection of signals onto null space of interference channels for avoiding secondary user interference to primary system has been proposed within the cognitive radio context. The idea is to estimate the null space of the channel matrix, between secondary user (SU) and primary user (PU). This process is done at the SU transmitter. In the case when channel is assumed to be reciprocal, between PU transmitter and receiver, null space can be estimated by SU transmitter by using second order statistics of the PU's transmitted signal [37, 38]. However, the above approach is restricted to PUs using Time Division Duplexing (TDD), for example, WiMax systems. If the assumption of channel reciprocity is removed, the channel estimation requires cooperation between the PU and the SU.

In this section, we formulate a projection algorithm to project the radar signal onto the null space of interference channel $\mathbf{H}_i$. Assuming, the MIMO radar has ICSI for all $\mathbf{H}_i$ interference channels, either through feedback or channel reciprocity, we can perform a singular value decomposition (SVD) to find the null space of $\mathbf{H}_i$ and use it to construct a projector matrix. First, we find SVD of $\mathbf{H}_i$, i.e.,

$$\mathbf{H}_i = \mathbf{U}_i \boldsymbol{\Sigma}_i \mathbf{V}_i^H. \tag{1.3}$$

Now, let us define

$$\widetilde{\boldsymbol{\Sigma}}_i \triangleq \operatorname{diag}(\widetilde{\sigma}_{i,1}, \widetilde{\sigma}_{i,2}, \dots, \widetilde{\sigma}_{i,p}) \tag{1.4}$$

where $p \triangleq \min(N^{\mathrm{BS}}, M)$ and $\widetilde{\sigma}_{i,1} > \widetilde{\sigma}_{i,2} > \cdots > \widetilde{\sigma}_{i,q} > \widetilde{\sigma}_{i,q+1} = \widetilde{\sigma}_{i,q+2} = \cdots \widetilde{\sigma}_{i,p} = 0$. Next, we define

$$\widetilde{\boldsymbol{\Sigma}}_i' \triangleq \operatorname{diag}(\widetilde{\sigma}_{i,1}', \widetilde{\sigma}_{i,2}', \dots, \widetilde{\sigma}_{i,M}') \tag{1.5}$$

where

$$\widetilde{\sigma}_{i,u}' \triangleq \begin{cases} 0, & \text{for } u \leqslant q, \\ 1, & \text{for } u > q. \end{cases} \tag{1.6}$$

Using above definitions we can now define our projection matrix, i.e.,

$$\mathbf{P}_i \triangleq \mathbf{V}_i \widetilde{\boldsymbol{\Sigma}}_i' \mathbf{V}_i^H. \tag{1.7}$$

Below, we show two properties of projection matrices showing that $\mathbf{P}_i$ is a valid projection matrix.

Property 1.1 $\mathbf{P}_i \in \mathbb{C}^{M\times M}$ *is a projection matrix if and only if* $\mathbf{P}_i = \mathbf{P}_i^H = \mathbf{P}_i^2$.

Proof Let's start by showing the 'only if' part. First, we show $\mathbf{P}_i = \mathbf{P}_i^H$. Taking Harmition of Eq. (1.7) we have

$$\mathbf{P}_i^H = (\mathbf{V}_i \widetilde{\boldsymbol{\Sigma}}_i' \mathbf{V}^H)^H = \mathbf{P}_i. \tag{1.8}$$

Now, squaring Eq. (1.7) we have

$$\mathbf{P}_i^2 = \mathbf{V}_i \widetilde{\boldsymbol{\Sigma}}_i \mathbf{V}^H \times \mathbf{V}_i \widetilde{\boldsymbol{\Sigma}}_i \mathbf{V}^H = \mathbf{P}_i \tag{1.9}$$

where above equation follows from $\mathbf{V}^H \mathbf{V}_i = \mathbf{I}$ (since they are orthonormal matrices) and $(\widetilde{\boldsymbol{\Sigma}}_i')^2 = \widetilde{\boldsymbol{\Sigma}}_i'$ (by construction). From Eqs. (1.8) and (1.9) it follows that $\mathbf{P}_i = \mathbf{P}_I^H = \mathbf{P}_i^2$. Next, we show $\mathbf{P}_i$ is a projector by showing that if $\mathbf{v} \in$ range $(\mathbf{P}_i)$, then $\mathbf{P}_i \mathbf{v} = \mathbf{v}$, i.e., for some $\mathbf{w}$, $\mathbf{v} = \mathbf{P}_i \mathbf{w}$, then

$$\mathbf{P}_i \mathbf{v} = \mathbf{P}_i (\mathbf{P}_i \mathbf{w}) = \mathbf{P}_i^2 \mathbf{w} = \mathbf{P}_i \mathbf{w} = \mathbf{v}. \tag{1.10}$$

Moreover, $\mathbf{P}_i \mathbf{v} - \mathbf{v} \in \text{null}(\mathbf{P}_i)$, i.e.,

$$\mathbf{P}_i (\mathbf{P}_i \mathbf{v} - \mathbf{v}) = \mathbf{P}_i^2 \mathbf{v} - \mathbf{P}_i \mathbf{v} = \mathbf{P}_i \mathbf{v} - \mathbf{P}_i \mathbf{v} = \mathbf{0}. \tag{1.11}$$

This concludes our proof.

Property 1.2 $\mathbf{P}_i \in \mathbb{C}^{M\times M}$ *is an orthogonal projection matrix onto the null space of* $\mathbf{H}_i \in \mathbb{C}^{N^{BS}\times M}$

Proof Since $\mathbf{P}_i = \mathbf{P}_i^H$, we can write

$$\mathbf{H}_i \mathbf{P}_i^H = \mathbf{U}_i \widetilde{\boldsymbol{\Sigma}}_i \mathbf{V}_i^H \times \mathbf{V}_i \widetilde{\boldsymbol{\Sigma}}_i' \mathbf{V}^H = \mathbf{0}. \tag{1.12}$$

The above results follows from noting that $\widetilde{\boldsymbol{\Sigma}}_i \widetilde{\boldsymbol{\Sigma}}_i' = \mathbf{0}$ by construction.

The formation of projection matrix for the BPSK and QPSK waveform design process is presented in the form of Algorithm 1.

Algorithm 1 Projection Algorithm

if $\mathbf{H}_i$ received from waveform design algorithm **then**
  Perform SVD on $\mathbf{H}_i$ (i.e. $\mathbf{H}_i = \mathbf{U}_i \boldsymbol{\Sigma}_i \mathbf{V}_i^H$)
  Construct $\widetilde{\boldsymbol{\Sigma}}_i = \text{diag}(\widetilde{\sigma}_{i,1}, \widetilde{\sigma}_{i,2}, \ldots, \widetilde{\sigma}_{i,p})$
  Construct $\widetilde{\boldsymbol{\Sigma}}_i' = \text{diag}(\widetilde{\sigma}_{i,1}', \widetilde{\sigma}_{i,2}', \ldots, \widetilde{\sigma}_{i,M}')$
  Setup projection matrix $\mathbf{P}_i = \mathbf{V}_i \widetilde{\boldsymbol{\Sigma}}_i' \mathbf{V}_i^H$.
  Send $\mathbf{P}_i$ to waveform design algorithm.
end if

References

1. J. Li, P. Stoica, *MIMO Radar Signal Processing* (Wiley-IEEE Press, 2008)
2. A.M. Haimovich, R.S. Blum, L.J. Cimini, MIMO radar with widely separated antennas. IEEE Signal Process. Mag. **25**(1), 116–129 (2008)
3. J. Li, P. Stoica, MIMO radar with colocated antennas. IEEE Signal Process. Mag. **24**(5), 106–114 (2007)
4. J. Tan, G. Stuber, Constant envelope multi-carrier modulation, in *Proceedings MILCOM 2002*, Oct 2002, vol. 1, pp. 607–611
5. S. Thompson, A. Ahmed, J. Proakis, J. Zeidler, M. Geile, Constant envelope OFDM. IEEE Trans. Commun. **56**, 1300–1312 (2008)
6. P. Stoica, J. Li, X. Zhu, Waveform synthesis for diversity-based transmit beampattern design. IEEE Trans. Signal Process. **56**, 2593–2598 (2008)
7. S. Ahmed, J.S. Thompson, Y.R. Petillot, B. Mulgrew, Finite alphabet constant-envelope waveform design for MIMO radar. IEEE Trans. Signal Process. **59**(11), 5326–5337 (2011)
8. S. Sodagari, A. Abdel-Hadi, Constant envelope radar with coexisting capability with LTE communication systems (under submission)
9. The Presidents Council of Advisors on Science and Technology (PCAST), Realizing the full potential of government-held spectrum to spur economic growth, July 2012
10. Federal Communications Commission (FCC), FCC proposes innovative small cell use in 3.5 GHz band. http://www.fcc.gov/document/fcc-proposes-innovative-small-cell-use-35-ghz-band. Accessed 12 Dec 2012
11. M. Bell, N. Devroye, D. Erricolo, T. Koduri, S. Rao, D. Tuninetti, Results on spectrum sharing between a radar and a communications system, in *International Conference on Electromagnetics in Advanced Applications (ICEAA)*, Aug 2014, pp. 826–829
12. D. Erricolo, H. Griffiths, L. Teng, M. Wicks, L. Lo Monte, On the spectrum sharing between radar and communication systems, in *2014 International Conference on Electromagnetics in Advanced Applications (ICEAA)*, pp. 890–893 (2014)
13. H. Shajaiah, A. Khawar, A. Abdel-Hadi, T.C. Clancy, Using resource allocation with carrier aggregation for spectrum sharing between radar and 4G-LTE cellular system, in *IEEE DySPAN* (2014)
14. M. Ghorbanzadeh, A. Abdelhadi, C. Clancy, A utility proportional fairness resource allocation in spectrally radar-coexistent cellular networks, in *Military Communications Conference (MILCOM)* (2014)
15. A. Khawar, A. Abdel-Hadi, T.C. Clancy, On the impact of time-varying interference-channel on the spatial approach of spectrum sharing between S-band radar and communication system, in *IEEE MILCOM* (2014)
16. National Telecommunications and Information Administration (NTIA), Analysis and resolution of RF interference to radars operating in the band 2700–2900 MHz from broadband communication transmitters, Oct 2012
17. A. Khawar, A. Abdelhadi, T. Clancy, A mathematical analysis of cellular interference on the performance of S-band military radar systems, in *Wireless Telecommunications Symposium*, Apr 2014
18. Federal Communications Commission (FCC), In the matter of revision of parts 2 and 15 of the commissions rules to permit unlicensed national information infrastructure (U-NII) devices in the 5 GHz band. MO&O, ET Docket No. 03–122, June 2006
19. National Telecommunications and Information Administration (NTIA), An assessment of the near-term viability of accommodating wireless broadband systems in the 1675–1710 MHz, 1755–1780 MHz, 3500–3650 MHz, 4200–4220 MHz, and 4380–4400 MHz bands (Fast Track Report), Oct 2010
20. A. Lackpour, M. Luddy, J. Winters, Overview of interference mitigation techniques between WiMAX networks and ground based radar, in *Proceedings of 20th Annual Wireless and Optical Communications Conference* (2011)

21. R. Saruthirathanaworakun, J. Peha, L. Correia, Performance of data services in cellular networks sharing spectrum with a single rotating radar, in *IEEE International Symposium on a World of Wireless, Mobile and Multimedia Networks (WoWMoM)* (2012), pp. 1–6
22. H. Deng, B. Himed, Interference mitigation processing for spectrum-sharing between radar and wireless communications systems. IEEE Trans. Aerosp. Electron. Syst. **49**(3), 1911–1919 (2013)
23. S. Sodagari, A. Khawar, T.C. Clancy, R. McGwier, A projection based approach for radar and telecommunication systems coexistence, in *IEEE Global Communications Conference (GLOBECOM)* (2012)
24. A. Khawar, A. Abdel-Hadi, T.C. Clancy, Spectrum sharing between S-band radar and LTE cellular system: A spatial approach, in *IEEE DySPAN* (2014)
25. D. Fuhrmann, G. San Antonio, Transmit beamforming for MIMO radar systems using signal cross-correlation. IEEE Trans. Aerosp. Electron. Syst. **44**, 171–186 (2008)
26. T. Aittomaki, V. Koivunen, Signal covariance matrix optimization for transmit beamforming in MIMO radars, in *Proceedings of the Forty-First Asilomar Conference on Signals, Systems and Computers (ASILOMAR)*, Nov 2007, pp. 182–186
27. P. Gong, Z. Shao, G. Tu, Q. Chen, Transmit beampattern design based on convex optimization for MIMO radar systems. Signal Process. **94**, 195–201 (2014)
28. G. Hua, S. Abeysekera, MIMO radar transmit beampattern design with ripple and transition band control. IEEE Trans. Signal Process. **61**, 2963–2974 (2013)
29. S. Jardak, S. Ahmed, M.-S. Alouini, Generating correlated QPSK waveforms by exploiting real Gaussian random variables, in *Forty Sixth Asilomar Conference on Signals, Systems and Computers (ASILOMAR)*, Nov 2012, pp. 1323–1327
30. A. Khawar, A. Abdelhadi, T.C. Clancy, QPSK waveform for MIMO radar with spectrum sharing constraints. arXiv:1407.8510
31. A. Khawar, A. Abdel-Hadi, T.C. Clancy, R. McGwier, Beampattern analysis for MIMO radar and telecommunication system coexistence, in *IEEE ICNC* (2014)
32. A. Khawar, A. Abdel-Hadi, T.C. Clancy, MIMO radar waveform design for coexistence with cellular systems, in *IEEE DySPAN* (2014)
33. A. Khawar, A. Abdelhadi, T.C. Clancy, Channel modeling between MIMO seaborne radar and MIMO cellular system. arXiv:1504.04325 (2015)
34. A. Khawar, A. Abdelhadi, T.C. Clancy, Performance analysis of coexisting radar and cellular system in LoS channel (under submission)
35. A. Khawar, A. Abdelhadi, T.C. Clancy, 3D channel modeling between seaborne radar and cellular system. arXiv:1504.04333 (2015)
36. D. Tse, P. Viswanath, *Fundamentals of Wireless Communication* (Cambridge University Press, 2005)
37. H. Yi, H. Hu, Y. Rui, K. Guo, J. Zhang, Null space-based precoding scheme for secondary transmission in a cognitive radio MIMO system using second-order statistics, in *Proceedings of IEEE International Conference on Communications ICC'09* (2009), pp. 1–5
38. H. Yi, Nullspace-based secondary joint transceiver scheme for cognitive radio MIMO networks using second-order statistics, in *IEEE International Communications Conference (ICC)* (2010)

Chapter 2
BPSK Radar Waveform

In this chapter, BPSK radar waveforms for spectrum sharing are designed, i.e., the problem of designing MIMO radar BPSK waveform to match a given beampattern in the presence of a cellular system is considered. The classical problem of beampattern matching is modified to include the constraint that the designed waveform should not cause interference to cellular system. So in addition to maximizing the received power at a number of given target locations and minimizing at all other locations this work also seeks to null out interference to cellular system through waveform design. The problem of waveform design for MIMO radars to coexist with a single communication system is considered in [1]. This work extends this approach and designs MIMO radar waveforms that can coexist with a cellular system, i.e., waveforms that support coexistence with many communication systems. Two types of radar platforms are considered. First, radar waveform is designed for stationary maritime MIMO radar that experiences stationary or slowly moving interference channel. Due to tractability of interference channel, null space projection (NSP) is included in unconstrained nonlinear optimization problem for waveform design. Second, radar waveform for moving maritime MIMO radar which experiences interference channel that is fast enough not to be included in optimization problem due to its intractability. For this case, FACE waveforms are designed first and then projected onto null space of interference channel before transmission. The performance of BPSK radar waveform for spectrum sharing is evaluated via numerical examples.

The remainder of this chapter is organized as follows. Section 2.1 discusses BPSK beampattern matching waveform design. Section 2.2 presents the synthesis of Gaussian covariance matrix for beampattern matching design problem. Section 2.3 solves the waveform design optimization problem for spectrum sharing. Section 2.4 discusses simulation setup and results. Section 2.5 concludes the chapter.

A. Khawar et al., *MIMO Radar Waveform Design for Spectrum Sharing with Cellular Systems*, SpringerBriefs in Electrical and Computer Engineering,
DOI 10.1007/978-3-319-29725-5_2

2.1 Finite Alphabet BPSK Beampattern Matching

In this section, finite alphabet BPSK waveforms are designed for spectrum sharing by considering a uniform linear array of M transmit antennas with inter-element spacing of half-wavelength. Then, the transmit signal is given as

$$\mathbf{x}(n) = \left[x_1(n)\ x_2(n)\ \cdots\ x_M(n)\right] \tag{2.1}$$

where $x_m(n)$ is the baseband signal from the mth transmit element at time index n. Then the received signal from a target at location θ_k is given as

$$r_k(n) = \sum_{m=1}^{M} e^{-j(m-1)\pi \sin\theta_k} x_m(n), \qquad k = 1, 2, \ldots, K. \tag{2.2}$$

The above received signal can be represented compactly as

$$r_k(n) = \mathbf{a}^H(\theta_k)\mathbf{x}(n) \tag{2.3}$$

where $\mathbf{a}(\theta_k)$ is the steering vector defined as

$$\mathbf{a}(\theta_k) = \left[1\ e^{-j\pi \sin\theta_k}\ e^{-j2\pi \sin\theta_k}\ \cdots\ e^{-j(M-1)\pi \sin\theta_k}\right] \tag{2.4}$$

Now, the power received from the target at location θ_k is given as

$$\begin{aligned} P(\theta_k) &= \mathbb{E}\{\mathbf{a}^H(\theta_k)\,\mathbf{x}(n)\,\mathbf{x}^H(n)\,\mathbf{a}(\theta_k)\} \\ &= \mathbf{a}^H(\theta_k)\,\mathbf{R}\,\mathbf{a}(\theta_k) \end{aligned} \tag{2.5}$$

where $\mathbf{R}$ is the correlation matrix of the transmitted signal. The desired beampattern $\phi(\theta_k)$ is formed by minimizing the square of the error between $P(\theta_k)$ and $\phi(\theta_k)$ through a cost function defined as

$$J(\mathbf{R}) = \frac{1}{K}\sum_{k=1}^{K}\left(\mathbf{a}^H(\theta_k)\,\mathbf{R}\,\mathbf{a}(\theta_k) - \phi(\theta_k)\right)^2. \tag{2.6}$$

It is important to realize that $\mathbf{R}$ can not be chosen freely since it is a covariance matrix of the transmitted waveform and thus it must be positive semidefinite. In addition, the interest is in constant envelope waveform, i.e., all antennas are required to transmit at same power level which translates to same diagonal elements of $\mathbf{R}$. Thus, $\mathbf{R}$ is subject to two constraints, namely,

$$\begin{aligned} &C_1 : \mathbf{v}^H\mathbf{R}\mathbf{v} \geqslant 0, && \forall\, \mathbf{v} \\ &C_2 : \mathbf{R}(m, m) = c, && m = 1, 2, \ldots, M. \end{aligned}$$

Thus, under the given constraints, a constrained nonlinear optimization problem can be setup to solve beampattern matching problem

$$\begin{aligned}
\min_{\mathbf{R}} \quad & \frac{1}{K}\sum_{k=1}^{K}\Big(\mathbf{a}^H(\theta_k)\,\mathbf{R}\,\mathbf{a}(\theta_k) - \phi(\theta_k)\Big)^2 \\
\text{subject to} \quad & \mathbf{v}^H\mathbf{R}\mathbf{v}, \qquad \forall\, \mathbf{v} \\
& \mathbf{R}(m,m) = c, \qquad m = 1, 2, \ldots, M.
\end{aligned} \tag{2.7}$$

For radar waveform design, this constrained nonlinear optimization problem can be transformed into an unconstrained nonlinear optimization problem by bounding the variables using multidimensional spherical coordinates [2]. Once $\mathbf{R}$ is synthesized, the waveform matrix $\mathbf{X}$ with N_s samples defined as

$$\mathbf{X} = \big[\mathbf{x}(1)\ \mathbf{x}(2)\ \cdots\ \mathbf{x}(N_s)\big]^T \tag{2.8}$$

can be realized from

$$\mathbf{X} = \mathcal{X}\boldsymbol{\Lambda}^{1/2}\mathbf{W}^H \tag{2.9}$$

where $\mathcal{X} \in \mathcal{C}^{N_s \times M}$ is a matrix of zero mean and unit variance Gaussian random variables, $\boldsymbol{\Lambda} \in \mathcal{R}^{M \times M}$ is the diagonal matrix of eigenvalues and $\mathbf{W} \in \mathcal{C}^{M \times M}$ is the matrix of eigenvectors of $\mathbf{R}$ [3]. Due to the distribution of $\mathcal{X}$, the distribution of the random variables in the columns of $\mathbf{X}$ is also Gaussian but the waveform produced is not guaranteed to have the CE property.

2.2 Gaussian Covariance Matrix Synthesis for Desired Beampattern

An algorithm to directly synthesize covariance matrix of Gaussian random variables to generate finite alphabet constant envelope binary phase shift keying (BPSK) waveform for a desired beampattern was proposed by Ahmed et al. [2]. Using the same approach, the Gaussian random variables with zero mean and unit variance, x_m, can be mapped onto BPSK symbol, z_m, through a simple relation

$$z_m = \text{sign}(x_m), \qquad m \in \{1, 2, \ldots, M\}. \tag{2.10}$$

Using results from [2], we have

$$\begin{aligned}
\mathbb{E}(z_p z_q) &= \mathbb{E}\Big(\text{sign}(x_p)\text{sign}(x_q)\Big) \\
&= \frac{2}{\pi}\sin^{-1}\big(\mathbb{E}(x_p x_q)\big)
\end{aligned} \tag{2.11}$$

where x_p and x_q are Gaussian random variables and z_p and z_q are BPSK random variables. Therefore, the relation between real covariance matrix of beampattern $\mathbf{R}$ and Gaussian covariance matrix $\mathbf{R}_g$ is given by

$$\mathbf{R} = \frac{2}{\pi} \sin^{-1}(\mathbf{R}_g). \tag{2.12}$$

The Gaussian covariance matrix $\mathbf{R}_g$ is generated by the matrix $\mathbf{X}$ of Gaussian random variables using (2.9). Then BPSK random variables are generated directly by

$$\mathbf{Z} = \text{sign}(\mathbf{X}). \tag{2.13}$$

In [3], the authors propose to synthesize $\mathbf{R}_g$ as $\mathbf{R}_g = \mathbf{U}^H\mathbf{U}$ which transforms Eq. (2.7) as

$$\min_{\psi_{ij}} \frac{1}{K} \sum_{k=1}^{K} \left(\frac{2}{\pi} \mathbf{a}^H(\theta_k) \sin^{-1}(\mathbf{U}^H\mathbf{U})\mathbf{a}(\theta_k) - \phi(\theta_k) \right)^2 \tag{2.14}$$

where ψ_{ij} are the variables of the optimization problem and $\mathbf{U}$ is given by

$$\mathbf{U} = \begin{pmatrix} 1 & \sin(\psi_{21}) & \sin(\psi_{31})\sin(\psi_{32}) & \cdots & \prod_{m=1}^{M-1} \sin(\psi_{Mm}) \\ 0 & \cos(\psi_{21}) & \sin(\psi_{31})\cos(\psi_{32}) & \cdots & \prod_{m=1}^{M-2} \sin(\psi_{Mm})\cos(\psi_{M,M-1}) \\ 0 & 0 & \cos(\psi_{31}) & \ddots & \vdots \\ \vdots & \vdots & \ddots & \cdots & \sin(\psi_{M1})\cos(\psi_{M2}) \\ 0 & 0 & \cdots & \cdots & \cos(\psi_{M1}) \end{pmatrix} \tag{2.15}$$

2.3 BPSK Waveform Design for Spectrum Sharing

This section considers the design of MIMO radar waveforms for spectrum sharing. Two waveform design approaches are considered: one includes spectrum sharing constraint in the optimization problem and the other does not. The motivation and reasons for these two approaches and their impact on radar waveform performance is discussed in the following sections.

We design MIMO radar waveform with the additional constraint of waveform being in null space of interference channel. In addition, we design the waveform without this constraint but project the designed waveform onto the null space of the interference channels.

The MIMO radar is sharing spectrum with a cellular system which has N_{BS} base stations, thus, there exist N_{BS} interference channels, i.e. $\mathbf{H}_i$, $i = 1, 2, \ldots, \mathcal{K}$, between the MIMO radar and the cellular system. In this chapter, we consider a MIMO radar that has less transmit antennas than the receive antennas of communication systems,

Algorithm 2 Interference-Channel-Selection Algorithm [4]

loop
  for $i = 1:$ **do**
    Estimate CSI of $\mathbf{H}_i$.
    Send $\mathbf{H}_i$ to Algorithm 3 for null space computation.
    Receive $\dim[\mathcal{N}(\mathbf{H}_i)]$ from Algorithm 3.
  end for
  Find $i_{\max} = \operatorname{argmax}_{1 \leqslant i \leqslant \mathcal{K}} \dim[\mathcal{N}(\mathbf{H}_i)]$.
  Set $\breve{\mathbf{H}} = \mathbf{H}_{i_{\max}}$ as the candidate interference channel.
  Send $\breve{\mathbf{H}}$ to Algorithm 3 to get NSP radar waveform.
end loop

i.e., $M \leqslant N_{\text{BS}}$. In this case, null space of interference channel is not well defined so we propose a threshold based scheme where singular values below a certain threshold will be selected to determine null space of interference channel. Thus, among $\mathcal{K}$ interference channels we select the interference channel for waveform design which has the maximum null space in order for the waveform to have optimum performance. For interference channel selection, two algorithms, Algorithms 2 and 3, are presented in [4] which first estimate the channel state information of interference channels. This is followed by the calculation of null space of interference channels and interference channel with the maximum null space is selected as the candidate channel. For our beampattern matching problem, we seek to select the best interference channel, defined as

$$\begin{aligned} i_{\max} &\triangleq \underset{1 \leqslant i \leqslant \mathcal{K}}{\operatorname{argmax}} \dim[\mathcal{N}(\mathbf{H}_i)] \\ \mathbf{H}_{\text{Best}} &\triangleq \mathbf{H}_{i_{\max}} \end{aligned}$$

and we seek to avoid the worst channel, defined as

$$\begin{aligned} i_{\min} &\triangleq \underset{1 \leqslant i \leqslant \mathcal{K}}{\operatorname{argmin}} \dim[\mathcal{N}(\mathbf{H}_i)] \\ \mathbf{H}_{\text{Worst}} &\triangleq \mathbf{H}_{i_{\min}} \end{aligned}$$

for MIMO radar waveform design.

Once $\mathbf{H}_{\text{Best}}$ or $\breve{\mathbf{H}}$ is selected the next step is to construct a projection matrix via singular value decomposition (SVD) theorem, which is given as

$$\begin{aligned} \breve{\mathbf{H}}^{N_{\text{BS}} \times M} &= \mathbf{U} \boldsymbol{\Sigma}^{N_{\text{BS}} \times M} \mathbf{V}^H \\ &= \mathbf{U} \begin{pmatrix} \sigma_1 & & & \\ & \sigma_2 & & 0 \\ & & \ddots & \\ & 0 & & \sigma_{j \in \min(N_{\text{BS}}, M)} \\ & & & \end{pmatrix} \mathbf{V}^H \end{aligned}$$

Algorithm 3 Modified NSP and Waveform Design

```
if H_i received from Algorithm 2 then
  Perform SVD on H_i (i.e. H_i = U_i Σ_i V_i^H)
  if σ_j ≠ 0 (i.e. jth singular value of Σ_i) then
    dim [N(H_i)] = 0
    Use pre-specified threshold δ
    for j = 1 : min(N_BS, M) do
      if σ_j < δ then
        dim [N(H_i)] = dim [N(H_i)] + 1
      else
        dim [N(H_i)] = 0
      end if
    end for
  else
    dim [N(H_i)] = The number of zero singular values
  end if
  Send dim [N(H_i)] to Algorithm 2.
end if
if Ȟ received from Algorithm 2 then
  Perform SVD on Ȟ = UΣV
  if σ_j ≠ 0 then
    Use pre-specified threshold δ
    σ_Null = {} {An empty set to collect σs below threshold δ}
    for j = 1 : min(N_BS, M) do
      if σ_j < δ then
        Add σ_j to σ_Null
      end if
    end for
    V̌ = σ_Null corresponding columns in V.
  end if
  Setup projection matrix P_V̌ = V̌V̌^H.
  Get NSP radar signal via Z_NSP = ZP_V̌^H.
end if
```

where $\mathbf{U}$ is the complex unitary matrix, $\boldsymbol{\Sigma}$ is the diagonal matrix of singular values, and $\mathbf{V}^H$ is the complex unitary matrix. If SVD results in non-zero singular values, we calculate null space numerically via Algorithm 3. A threshold is defined and all the vectors in $\mathbf{V}^H$ corresponding to singular values below the threshold are collected in $\check{\mathbf{V}}$. Then, the projection matrix is formulated as in [5, 6]

$$\mathbf{P}_{\check{\mathbf{V}}} = \check{\mathbf{V}}\check{\mathbf{V}}^H. \tag{2.16}$$

2.3.1 *BPSK Waveform for Stationary MIMO Radar*

Consider the case of a maritime MIMO radar when a ship is docked or is stationary and thus radar platform is stationary. In this case, interference channels have little to no variations and thus it is feasible to include the constraint of NSP into the optimization problem. The new optimization problem is formulated by the combination of projection matrix, Eq. (2.16), into the optimization problem in Eq. (2.14) as

$$\min_{\psi_{ij}} \frac{1}{K} \sum_{k=1}^{K} \left(\frac{2}{\pi} \mathbf{a}^H(\theta_k) \mathbf{P}_{\check{\mathbf{V}}} \sin^{-1}(\mathbf{U}^H \mathbf{U}) \mathbf{P}_{\check{\mathbf{V}}}^H \mathbf{a}(\theta_k) - \phi(\theta_k) \right)^2. \tag{2.17}$$

This optimization problem does not guarantee to generate constant envelope radar waveform but guarantees that the designed waveform is in the null space of the interference channel or the designed waveform does not cause interference to the communication system. In addition, it is an evaluation of the impact of the NSP on the CE radar waveforms. The waveform generation process is shown using the block diagram of Fig. 2.1. The waveform generated by solving the optimization problem in Eq. (2.17) and then using Eq. (2.9) is denoted by $\mathbf{X}_{\text{opt}}$. The corresponding BPSK waveform is denoted by $\mathbf{Z}_{\text{opt}}$ which is obtained using Eq. (2.13). Then, the output NSP waveform is given by

$$\mathbf{Z}_{\text{opt-NSP}} = \mathbf{Z}_{\text{opt}} \mathbf{P}_{\check{\mathbf{V}}}^H. \tag{2.18}$$

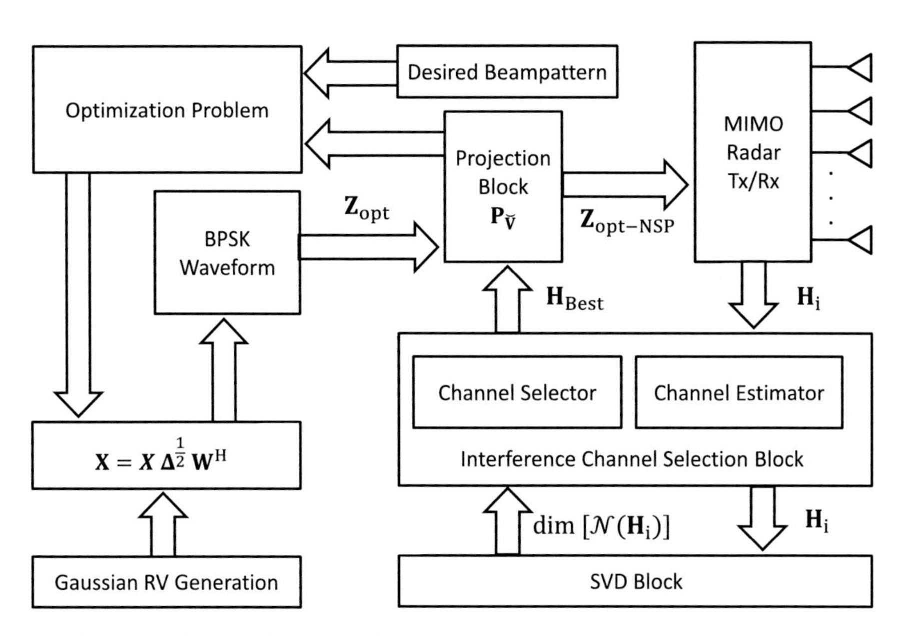

Fig. 2.1 Block diagram of the transmit beampattern design problem for a stationary maritime MIMO radar. The desired waveform is generated by including the projection matrix $\mathbf{P}_{\check{\mathbf{V}}}$, for the candidate interference channel $\mathbf{H}_{\text{Best}}$, in the optimization process. For this waveform constant envelope property is not guaranteed. The candidate interference channel is selected by Algorithms 2 and 3

2.3.2 *BPSK Waveform for Moving MIMO Radar*

Consider the case of a maritime radar which is moving and thus experiences interference channels that change too fast. In this case, it is not feasible to include the NSP in the optimization problem. Alternately, we can design CE waveforms by solving the optimization problem in Eq. (2.14) and then projection the waveform onto the null space of interference channel. The waveform generation process is shown using the block diagram of Fig. 2.2. Thus, the CE waveform is generated and then projected onto the null space of the interference channel according to

$$\mathbf{Z}_{\text{NSP}} = \mathbf{Z}\mathbf{P}_{\check{\mathbf{V}}}^{H}. \tag{2.19}$$

This formulation projects the CE waveform onto the null space of the interference channel. In the next section, the impact of projection on the radar waveform performance is studied.

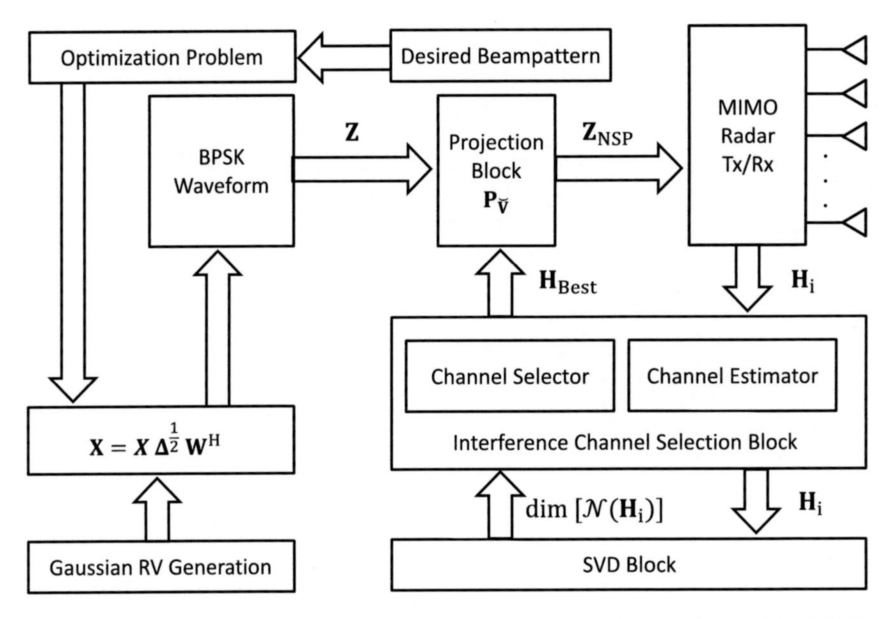

Fig. 2.2 Block diagram of the transmit beampattern design problem for a moving maritime MIMO radar. The desired waveform is generated with constant envelope property and then projected onto the candidate interference channel $\mathbf{H}_{\text{Best}}$ via projection matrix $\mathbf{P}_{\check{\mathbf{V}}}$. The candidate interference channel is selected by Algorithms 2 and 3

2.4 Numerical Examples

This section provides numerical examples discussing BPSK waveforms for spectrum sharing. A uniform linear array (ULA) of ten elements, i.e., $M = 10$, is considered with an interelement spacing of half-wavelength. In addition, all antennas transmit at the same power level which is fixed to unity. Each antenna transmits a waveform with $N_s = 100$ symbols and the resulting beampattern is the average of 100 Monte-Carlo trials of BPSK waveforms. The mean-squared error (MSE) between the desired, $\phi(\theta_k)$, and actual beampatterns, $P(\theta_k)$, is given by

$$\mathrm{MSE} = \frac{1}{K}\sum_{k=1}^{K}\Big(P(\theta_k) - \phi(\theta_k)\Big)^2.$$

The interference channels, $\mathbf{H}_i$, are modeled as a Rayleigh fading channels with Rayleigh probability density function (pdf) given by

$$f(h|\rho) = \frac{h}{\rho^2}e^{\frac{-h^2}{2\rho^2}}$$

where ρ is the mode of the Rayleigh distribution. The candidate interference channel, $\breve{\mathbf{H}}$, for waveform design is selected using Algorithm 2 and its null space is computed using SVD according to Algorithm 3.

In Fig. 2.3, the desired beampattern has two main lobes from $-60°$ to $-40°$ and from $40°$ to $60°$. It is the beampattern for stationary maritime MIMO radar obtained

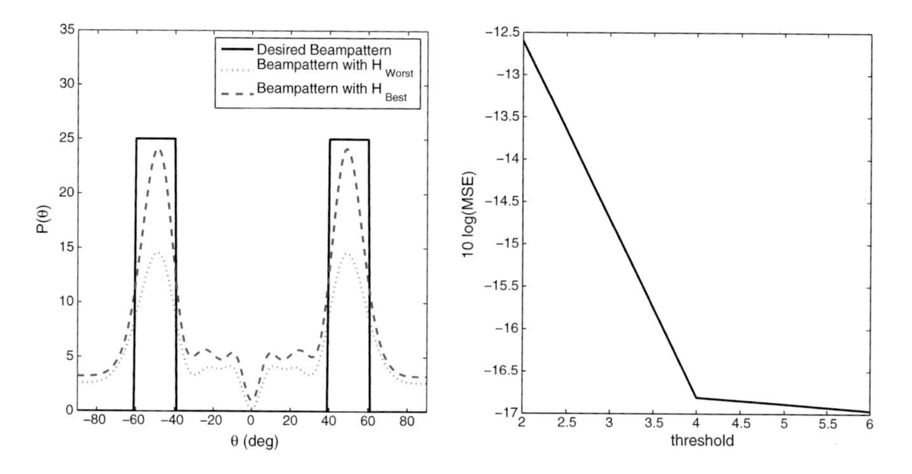

Fig. 2.3 Transmit beampattern and its MSE for a *stationary* maritime MIMO radar. The figure compares the desired beampattern with the average beampattern of BPSK waveforms for null space projection *included* in beampattern matching optimization problem for candidate interference channel $\mathbf{H}_{\text{Best}}$

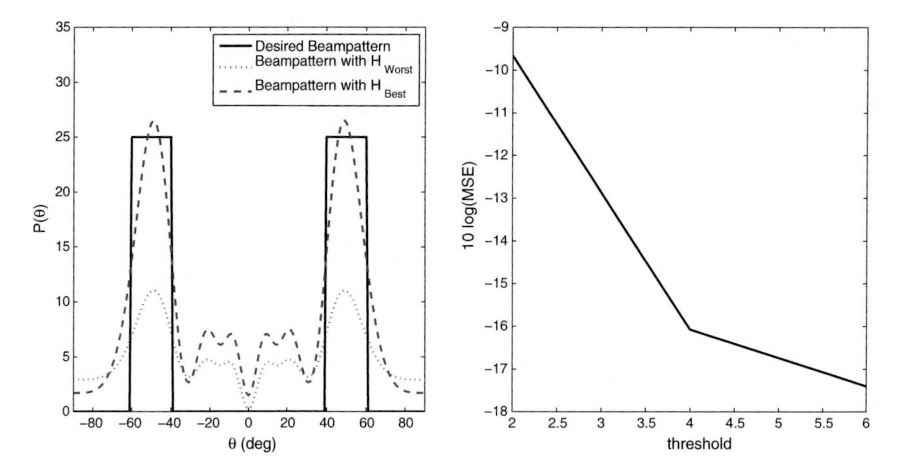

Fig. 2.4 Transmit beampattern and its MSE for a *moving* maritime MIMO radar. The figure compares the desired beampattern with the average beampattern of BPSK waveforms for null space projection *after* optimization for candidate interference channel $\mathbf{H}_{\text{Best}}$

by solving the optimization problem in Eq. (2.17). The resulting waveform covariance matrix is given by

$$\mathbf{R}_{\text{opt-NSP}} = \frac{1}{N_s}\mathbf{Z}^H_{\text{opt-NSP}}\mathbf{Z}_{\text{opt-NSP}}$$

Note that the desired beampattern and the beampattern obtained by including the projection matrix inside the optimization problem match closely for interference channel $\mathbf{H}_{\text{Best}}$ than $\mathbf{H}_{\text{Worst}}$.

In Fig. 2.4, the desired beampattern has two main lobes from $-60°$ to $-40°$ and from $40°$ to $60°$. It represents the beampattern of a moving maritime MIMO radar. Since interference channels are evolving fast, beampattern is obtained by solving the optimization problem in Eq. (2.14) and then projecting the resulting waveform onto the null space of $\breve{\mathbf{H}}$ using the projection matrix in Eq. (2.16). The resulting waveform covariance matrix is given by

$$\mathbf{R}_{\text{NSP}} = \frac{1}{N_s}\mathbf{Z}^H_{\text{NSP}}\mathbf{Z}_{\text{NSP}}.$$

Note that the desired beampattern and the beampattern obtained by projecting the waveform onto the null space of interference channel match closely for interference channel $\mathbf{H}_{\text{Best}}$ than $\mathbf{H}_{\text{Worst}}$.

In Figs. 2.3 and 2.4, MSE of beampattern matching design problem is shown. It shows that interference channel with the largest null space have the least MSE. This is in accordance with the methodology to select $\mathbf{H}_{\text{Best}}$ among k interference channels using Algorithms 2 and 3.

In Figs. 2.3 and 2.4, the desired beampattern match closely the actual beampattern when interference channel $\mathbf{H}_{\mathrm{Best}}$ is used. Thus, by careful selection of interference channel using Algorithms 2 and 3, when sharing spectrum with a cellular system, we can obtain a beampattern which is very close to the desired beampattern and in addition do not interfere with the communication system

2.5 Conclusion

In this chapter, we considered the MIMO radar waveform design from a spectrum sharing perspective. We considered a MIMO radar and a cellular system sharing spectrum and we designed radar waveforms such that they are not interfering with the cellular system. A method to design MIMO radar waveforms was presented which matched the beampattern to a certain desired beam pattern with the constraints that the waveform should have constant envelope and belong to the null space of interference channel. We designed waveform for the case when the MIMO radar is stationary and thus NSP can be included in the optimization problem due to the tractability of interference channel. We also designed waveform for the case when the MIMO radar is moving and experiences rapidly changing interference channels. This problem didn't consider the inclusion of NSP in the optimization problem due to the intractability of interference channel but rather constructed a CE radar waveform and projected it onto null space of interference channel. The interference channel was selected using Algorithms 2 and 3 and results showed that for both type of waveforms the desired beampattern and NSP beampatterns matched closely.

References

1. S. Sodagari, A. Abdel-Hadi, Constant envelope radar with coexisting capability with LTE communication systems (under submission)
2. S. Ahmed, J. Thompson, Y. Petillot, B. Mulgrew, Unconstrained synthesis of covariance matrix for MIMO radar transmit beampattern. IEEE Trans. Signal Process. **59**, 3837–3849 (2011)
3. S. Ahmed, J.S. Thompson, Y.R. Petillot, B. Mulgrew, Finite alphabet constant-envelope waveform design for MIMO radar. IEEE Trans. Signal Process. **59**(11), 5326–5337 (2011)
4. A. Khawar, A. Abdel-Hadi, T.C. Clancy, Spectrum sharing between S-band radar and LTE cellular system: A spatial approach, in *IEEE DySPAN* (2014)
5. S. Sodagari, A. Khawar, T.C. Clancy, R. McGwier, A projection based approach for radar and telecommunication systems coexistence, in *IEEE Global Communications Conference (GLOBECOM)* (2012)
6. A. Khawar, A. Abdel-Hadi, T.C. Clancy, R. McGwier, Beampattern analysis for MIMO radar and telecommunication system coexistence, in *IEEE ICNC* (2014)

Chapter 3
QPSK Radar Waveform

Multiple-input multiple-output (MIMO) radar is a relatively new concept in the field of radar signal processing. Many novel MIMO radar waveforms have been developed by considering various performance metrics and constraints. In this chapter, we show that finite alphabet constant-envelope (FACE) quadrature-pulse shift keying (QPSK) waveforms can be designed to realize a given covariance matrix by transforming a constrained nonlinear optimization problem into an unconstrained nonlinear optimization problem. In addition, we design QPSK waveforms in a way that they don't cause interference to cellular system, by steering nulls towards a selected base station (BS). The BS is selected according to our algorithm which guarantees minimum degradation in radar performance due to null space projection (NSP) of radar waveforms. We design QPSK waveforms with spectrum sharing constraints for stationary and moving radar platform. We show that the waveform designed for stationary MIMO radar matches the desired beampattern closely, when the number of BS antennas N_{BS} is considerably less than the number of radar antennas M, due to quasi-static interference channel. However, for moving radar the difference between designed and desired waveforms is larger than stationary radar, due to rapidly changing wireless channel.

This chapter is organized as follows. Section 3.1 introduces finite alphabet constant-envelope beampattern matching design problem. Section 3.2 introduces QPSK radar waveforms. Section 3.4 designs QPSK waveforms with spectrum sharing constraints for stationary and moving radar platforms. Section 3.5 discusses simulation setup and results. Section 3.6 concludes the chapter.

The content in this chapter is reproduced with permission after some modifications. For the original article please refer to: A. Khawar, A. Abdelhadi, T.C. Clancy, "QPSK waveform for MIMO radar with spectrum sharing constraints", Physical Communication, Vol 17, pg. 37–57, 2015.

A. Khawar et al., *MIMO Radar Waveform Design for Spectrum Sharing with Cellular Systems*, SpringerBriefs in Electrical and Computer Engineering,
DOI 10.1007/978-3-319-29725-5_3

3.1 Finite Alphabet Constant-Envelope Beampattern Design

In this chapter, we design QPSK waveforms having finite alphabets and constant-envelope property. We consider a uniform linear array (ULA) of M transmit antennas with inter-element spacing of half-wavelength. Then, the transmitted QPSK signal is given as

$$\widetilde{\mathbf{x}}(n) = \left[\widetilde{x}_1(n)\ \widetilde{x}_2(n)\ \cdots\ \widetilde{x}_M(n)\right]^T \tag{3.1}$$

where $\widetilde{x}_m(n)$ is the QPSK signal from the mth transmit element at time index n. Then, the received signal from a target at location θ_k is given as

$$\widetilde{r}_k(n) = \sum_{m=1}^{M} e^{-j(m-1)\pi \sin\theta_k} \widetilde{x}_m(n), \quad k = 1, 2, \ldots, K, \tag{3.2}$$

where K is the total number of targets. We can write the received signal compactly as

$$\widetilde{r}_k(n) = \mathbf{a}^H(\theta_k)\widetilde{\mathbf{x}}(n) \tag{3.3}$$

where $\mathbf{a}(\theta_k)$ is the steering vector defined as

$$\mathbf{a}(\theta_k) = \left[1\ e^{-j\pi \sin\theta_k}\ \ldots\ e^{-j(M-1)\pi \sin\theta_k}\right]^T. \tag{3.4}$$

We can write the power received at the target located at θ_k as

$$\begin{aligned} P(\theta_k) &= \mathbb{E}\{\mathbf{a}^H(\theta_k)\,\widetilde{\mathbf{x}}(n)\,\widetilde{\mathbf{x}}^H(n)\,\mathbf{a}(\theta_k)\} \\ &= \mathbf{a}^H(\theta_k)\,\widetilde{\mathbf{R}}\,\mathbf{a}(\theta_k) \end{aligned} \tag{3.5}$$

where $\widetilde{\mathbf{R}}$ is correlation matrix of the transmitted QPSK waveform. The desired QPSK beampattern $\phi(\theta_k)$, which represents the desired power at location θ_k, is formed by minimizing the square of the error between $P(\theta_k)$ and $\phi(\theta_k)$ through a cost function defined as

$$J(\widetilde{\mathbf{R}}) = \frac{1}{K}\sum_{k=1}^{K}\left(\mathbf{a}^H(\theta_k)\,\widetilde{\mathbf{R}}\,\mathbf{a}(\theta_k) - \phi(\theta_k)\right)^2. \tag{3.6}$$

Since, $\widetilde{\mathbf{R}}$ is covariance matrix of the transmitted signal it must be positive semi-definite. Moreover, due to the interest in constant-envelope property of waveforms, all antennas must transmit at the same power level. The optimization problem in Eq. (3.6) has some constraints and, thus, can't be chosen freely. In order to design finite alphabet constant-envelope waveforms, we must satisfy the following constraints:

$$\begin{aligned} &C_1 : \mathbf{v}^H\widetilde{\mathbf{R}}\mathbf{v} \geqslant 0, && \forall\, \mathbf{v}, \\ &C_2 : \widetilde{\mathbf{R}}(m, m) = c, && m = 1, 2, \ldots, M, \end{aligned}$$

where C_1 satisfies the 'positive semi-definite' constraint and C_2 satisfies the 'constant-envelope' constraint. Thus, we have a constrained nonlinear optimization problem given as

$$\begin{array}{lll} \min\limits_{\widetilde{\mathbf{R}}} & \frac{1}{K}\sum_{k=1}^{K}\left(\mathbf{a}^H(\theta_k)\,\widetilde{\mathbf{R}}\,\mathbf{a}(\theta_k) - \phi(\theta_k)\right)^2 & \\ \text{subject to} & \mathbf{v}^H\widetilde{\mathbf{R}}\mathbf{v} \geqslant 0, & \forall\, \mathbf{v}, \\ & \widetilde{\mathbf{R}}(m,m) = c, & m = 1, 2, \ldots, M. \end{array} \tag{3.7}$$

Ahmed et al. showed that, by using multi-dimensional spherical coordinates, this constrained nonlinear optimization can be transformed into an unconstrained nonlinear optimization [1]. Once $\widetilde{\mathbf{R}}$ is synthesized, the waveform matrix $\widetilde{\mathbf{X}}$ with N samples is given as

$$\widetilde{\mathbf{X}} = \left[\widetilde{\mathbf{x}}(1)\ \widetilde{\mathbf{x}}(2)\ \cdots\ \widetilde{\mathbf{x}}(N)\right]^T. \tag{3.8}$$

This can be realized from

$$\widetilde{\mathbf{X}} = \mathcal{X}\boldsymbol{\Lambda}^{1/2}\mathbf{W}^H \tag{3.9}$$

where $\mathcal{X} \in \mathcal{C}^{N\times M}$ is a matrix of zero mean and unit variance Gaussian random variables, $\boldsymbol{\Lambda} \in \mathcal{R}^{M\times M}$ is the diagonal matrix of eigenvalues, and $\mathbf{W} \in \mathcal{C}^{M\times M}$ is the matrix of eigenvectors of $\widetilde{\mathbf{R}}$ [2]. Note that $\widetilde{\mathbf{X}}$ has Gaussian distribution due to $\mathcal{X}$ but the waveform produced is not guaranteed to have the CE property.

3.2 Finite Alphabet Constant-Envelope QPSK Waveforms

Consider zero mean and unit variance Gaussian random variables (RVs) $\widetilde{x}_m$ and $\widetilde{y}_m$ that can be mapped onto a QPSK RV $\widetilde{z}_m$ through, as in [3],

$$\widetilde{z}_m = \frac{1}{\sqrt{2}}\left[\operatorname{sign}(\widetilde{x}_m) + \jmath\ \operatorname{sign}(\widetilde{y}_m)\right]. \tag{3.10}$$

Then, it is straight forward to write the (p, q)th element of the complex covariance matrix as

$$\mathbb{E}\{\widetilde{z}_p\widetilde{z}_q\} = \gamma_{pq} = \gamma_{\Re_{pq}} + \jmath\,\gamma_{\Im_{pq}} \tag{3.11}$$

where $\gamma_{\Re_{pq}}$ and $\gamma_{\Im_{pq}}$ are the real and imaginary parts of γ_{pq}, respectively. If, Gaussian RVs $\widetilde{x}_p$, $\widetilde{x}_q$, $\widetilde{y}_p$, and $\widetilde{y}_q$ are chosen such that

$$\begin{aligned} \mathbb{E}\{\widetilde{x}_p\widetilde{x}_q\} &= \mathbb{E}\{\widetilde{y}_p\widetilde{y}_q\} \\ \mathbb{E}\{\widetilde{x}_p\widetilde{y}_q\} &= -\mathbb{E}\{\widetilde{y}_p\widetilde{x}_q\} \end{aligned} \tag{3.12}$$

then we can write the real and imaginary parts of γ_{pq} as

$$\gamma_{\Re_{pq}} = \mathbb{E}\Big\{\text{sign}(\widetilde{x}_p)\text{sign}(\widetilde{x}_q)\Big\}$$
$$\gamma_{\Im_{pq}} = \mathbb{E}\Big\{\text{sign}(\widetilde{y}_p)\text{sign}(\widetilde{x}_q)\Big\}. \tag{3.13}$$

Then, from Eq. (A.18) Appendix A.2, we have

$$\mathbb{E}\{\widetilde{z}_p\widetilde{z}_q\} = \frac{2}{\pi}\Bigg[\sin^{-1}\Big(\mathbb{E}\{\widetilde{x}_p\widetilde{x}_q\}\Big) + \jmath\ \sin^{-1}\Big(\mathbb{E}\{\widetilde{y}_p\widetilde{x}_q\}\Big)\Bigg]. \tag{3.14}$$

The complex Gaussian covariance matrix $\widetilde{\mathbf{R}}_g$ is defined as

$$\widetilde{\mathbf{R}}_g \triangleq \Re(\mathbf{R}_g) + \jmath\,\Im(\mathbf{R}_g) \tag{3.15}$$

where $\Re(\mathbf{R}_g)$ and $\Im(\mathbf{R}_g)$ both have real entries, since $\mathbf{R}_g$ is a real Gaussian covariance matrix. Then, Eq. (3.14) can be written as

$$\widetilde{\mathbf{R}} = \frac{2}{\pi}\Bigg[\sin^{-1}\Big(\Re(\mathbf{R}_g)\Big) + \jmath\ \sin^{-1}\Big(\Im(\mathbf{R}_g)\Big)\Bigg]. \tag{3.16}$$

In [3], it is proposed to construct complex Gaussian covariance matrix via transform $\widetilde{\mathbf{R}}_g = \widetilde{\mathbf{U}}^H\widetilde{\mathbf{U}}$, where $\widetilde{\mathbf{U}}$ is given by Eq. (3.19). Then, $\widetilde{\mathbf{U}}$ can be written as

$$\widetilde{\mathbf{U}} = \Re(\widetilde{\mathbf{U}}) + \jmath\Im(\widetilde{\mathbf{U}}) \tag{3.17}$$

where $\Re(\widetilde{\mathbf{U}})$ and $\Im(\widetilde{\mathbf{U}})$ are given by Eqs. (3.20) and (3.21), respectively. Alternately, $\widetilde{\mathbf{R}}_g$ can also be expressed as

$$\widetilde{\mathbf{R}}_g = \Big[\Re(\widetilde{\mathbf{U}})^H\Re(\widetilde{\mathbf{U}}) + \Im(\widetilde{\mathbf{U}})^H\Im(\widetilde{\mathbf{U}})\Big] + \jmath\,\Big[\Re(\widetilde{\mathbf{U}})^H\Im(\widetilde{\mathbf{U}}) - \Im(\widetilde{\mathbf{U}})^H\Re(\widetilde{\mathbf{U}})\Big]. \tag{3.18}$$

$$\widetilde{\mathbf{U}} = \begin{pmatrix} e^{j\psi_1} & e^{j\psi_2}\sin(\psi_{21}) & e^{j\psi_3}\sin(\psi_{31})\sin(\psi_{32}) & \cdots & e^{j\psi_M}\prod_{m=1}^{M-1}\sin(\psi_{Mm}) \\ 0 & e^{j\psi_2}\cos(\psi_{21}) & e^{j\psi_3}\sin(\psi_{31})\cos(\psi_{32}) & \cdots & e^{j\psi_M}\prod_{m=1}^{M-2}\sin(\psi_{Mm})\cos(\psi_{M,M-1}) \\ 0 & 0 & e^{j\psi_3}\cos(\psi_{31}) & \ddots & \vdots \\ \vdots & \vdots & \ddots & \cdots & e^{j\psi_M}\sin(\psi_{M1})\cos(\psi_{M2}) \\ 0 & 0 & \cdots & \cdots & e^{j\psi_M}\cos(\psi_{M1}) \end{pmatrix} \tag{3.19}$$

$$\Re(\widetilde{\mathbf{U}}) = \begin{pmatrix} \cos(\psi_1) & \cos(\psi_2)\sin(\psi_{21}) & \cos(\psi_3)\sin(\psi_{31})\sin(\psi_{32}) & \cdots & \cos(\psi_M)\prod_{m=1}^{M-1}\sin(\psi_{Mm}) \\ 0 & \cos(\psi_2)\cos(\psi_{21}) & \cos(\psi_3)\sin(\psi_{31})\cos(\psi_{32}) & \cdots & \cos(\psi_M)\prod_{m=1}^{M-2}\sin(\psi_{Mm})\cos(\psi_{M,M-1}) \\ 0 & 0 & \cos(\psi_3)\cos(\psi_{31}) & \ddots & \vdots \\ \vdots & \vdots & \ddots & \cdots & \cos(\psi_M)\sin(\psi_{M1})\cos(\psi_{M2}) \\ 0 & 0 & \cdots & \cdots & \cos(\psi_M)\cos(\psi_{M1}) \end{pmatrix} \tag{3.20}$$

$$\Im(\widetilde{\mathbf{U}}) = \begin{pmatrix} \sin(\psi_1) & \sin(\psi_2)\sin(\psi_{21}) & \sin(\psi_3)\sin(\psi_{31})\sin(\psi_{32}) & \cdots & \sin(\psi_M)\prod_{m=1}^{M-1}\sin(\psi_{Mm}) \\ 0 & \sin(\psi_2)\cos(\psi_{21}) & \sin(\psi_3)\sin(\psi_{31})\cos(\psi_{32}) & \cdots & \sin(\psi_M)\prod_{m=1}^{M-2}\sin(\psi_{Mm})\cos(\psi_{M,M-1}) \\ 0 & 0 & \sin(\psi_3)\cos(\psi_{31}) & \ddots & \vdots \\ \vdots & \vdots & \ddots & \cdots & \sin(\psi_M)\sin(\psi_{M1})\cos(\psi_{M2}) \\ 0 & 0 & \cdots & \cdots & \sin(\psi_M)\cos(\psi_{M1}) \end{pmatrix} \tag{3.21}$$

Lemma 3.1 *If $\mathbf{R}_g$ is a covariance matrix and*

$$\widetilde{\mathbf{R}}_g = \Re(\mathbf{R}_g) + \jmath\,\Im(\mathbf{R}_g) \tag{3.22}$$

then the complex covariance matrix $\widetilde{\mathbf{R}}_g$ will always be positive semi-definite.

Proof Please see Appendix A.3.

Lemma 3.1 satisfies constraint C_1 and $\widetilde{\mathbf{R}}_g$ also satisfies constraint C_2 for $c = 1$. This helps to transform constrained nonlinear optimization into unconstrained nonlinear optimization in the following section.

In order to generate QPSK waveforms we define $N \times 2M$ matrix $\widetilde{\mathbf{S}}$, of Gaussian RVs, as

$$\widetilde{\mathbf{S}} \triangleq \left[\widetilde{\mathbf{X}}\ \widetilde{\mathbf{Y}}\right] \tag{3.23}$$

where $\widetilde{\mathbf{X}}$ and $\widetilde{\mathbf{Y}}$ are of each size $N \times M$, representing real and imaginary parts of QPSK waveform matrix, which is given as

$$\widetilde{\mathbf{Z}} = \frac{1}{\sqrt{2}}\left[\operatorname{sign}(\widetilde{\mathbf{X}}) + \jmath\,\operatorname{sign}(\widetilde{\mathbf{Y}})\right]. \tag{3.24}$$

The covariance matrix of $\widetilde{\mathbf{S}}$ is given as

$$\widetilde{\mathbf{R}}_{\widetilde{\mathbf{S}}} = \mathbb{E}\{\widetilde{\mathbf{S}}^H\widetilde{\mathbf{S}}\} = \begin{bmatrix} \Re(\mathbf{R}_g) & \Im(\mathbf{R}_g) \\ -\Im(\mathbf{R}_g) & \Re(\mathbf{R}_g) \end{bmatrix}. \tag{3.25}$$

QPSK waveform matrix $\widetilde{\mathbf{Z}}$ can be realized by the matrix $\widetilde{\mathbf{S}}$ of Gaussian RVs which can be generated using Eq. (3.9) by utilizing $\widetilde{\mathbf{R}}_{\widetilde{\mathbf{S}}}$.

3.3 Gaussian Covariance Matrix Synthesis for Desired Beampattern

In this section, we prove that the desired QPSK beampattern can be directly synthesized by using the complex covariance matrix, $\widetilde{\mathbf{R}}_g$, for complex Gaussian RVs. This generates M QPSK waveforms for the desired beampattern which satisfy the property of finite alphabet and constant-envelope. By exploiting the relationship between the complex Gaussian RVs and QPSK RVs we have

$$\widetilde{\mathbf{R}} = \frac{2}{\pi}\Big[\sin^{-1}\Big(\Re(\mathbf{R}_g)\Big) + \jmath \sin^{-1}\Big(\Im(\mathbf{R}_g)\Big)\Big]. \tag{3.26}$$

Lemma 3.2 *If $\widetilde{\mathbf{R}}_g$ is a complex covariance matrix and*

$$\widetilde{\mathbf{R}} = \frac{2}{\pi}\Big[\sin^{-1}\Big(\Re(\mathbf{R}_g)\Big) + \jmath \sin^{-1}\Big(\Im(\mathbf{R}_g)\Big)\Big]$$

then $\widetilde{\mathbf{R}}$ will always be positive semi-definite.

Proof Please see Appendix A.3.

Using Eq. (3.26) we can rewrite the optimization problem in Eq. (3.7) as

$$\begin{array}{ll} \min\limits_{\widetilde{\mathbf{R}}} & \frac{1}{K}\sum_{k=1}^{K}\Big[\frac{2}{\pi}\mathbf{a}^H(\theta_k)\Big\{\sin^{-1}\Big(\Re(\mathbf{R}_g)\Big) + \jmath \sin^{-1}\Big(\Im(\mathbf{R}_g)\Big)\Big\}\mathbf{a}(\theta_k) - \phi(\theta_k)\Big]^2 \\ \text{subject to} & \mathbf{v}^H\widetilde{\mathbf{R}}\mathbf{v} \geqslant 0, \qquad \forall\, \mathbf{v}, \\ & \widetilde{\mathbf{R}}(m,m) = c, \qquad m = 1, 2, \ldots, M. \end{array} \tag{3.27}$$

$$\begin{aligned} J(\boldsymbol{\Theta}) =& \frac{1}{K}\sum_{k=1}^{K}\Big[\frac{2}{\pi}\mathbf{a}^H(\theta_k)\Big\{\sin^{-1}\Big(\Re(\widetilde{\mathbf{U}})^H\Re(\widetilde{\mathbf{U}}) + \Im(\widetilde{\mathbf{U}})^H\Im(\widetilde{\mathbf{U}})\Big) \\ &+ \jmath \sin^{-1}\Big(\Re(\widetilde{\mathbf{U}})^H\Im(\widetilde{\mathbf{U}}) - \Im(\widetilde{\mathbf{U}})^H\Re(\widetilde{\mathbf{U}})\Big)\Big\}\mathbf{a}^H(\theta_k) - \alpha\phi(\theta_k)\Big]^2 \end{aligned} \tag{3.28}$$

Since, the matrix $\widetilde{\mathbf{U}}$ is already known, we can formulate $\widetilde{\mathbf{R}}_g$ via Eq. (3.18). We can also write the (p, q)th element of the upper triangular matrix $\widetilde{\mathbf{R}}_g$ by first writing the (p, q)th element of the upper triangular matrix $\Re\big(\mathbf{R}_g(p, q)\big)$ as

$$\Re\big(\mathbf{R}_g(p,q)\big) = \begin{cases} \prod_{l=1}^{q-1}\sin(\psi_{ql})\prod_{s=1}^{p}\prod_{u=1}^{q} f(s,u), & p > q \\ 1, & p = q \end{cases} \tag{3.29}$$

where $f(s, u) = \cos(\psi_s)\cos(\psi_u) + \sin(\psi_s)\sin(\psi_u)$; and the (p, q)th element of the upper triangular matrix $\Im\big(\mathbf{R}_g(p, q)\big)$ as

$$\Im\big(\mathbf{R}_g(p,q)\big) = \begin{cases} g(p,q)\prod_{l=1}^{q-1}\sin(\psi_{ql}), & p > q \\ 0, & p = q \end{cases} \tag{3.30}$$

where $g(p, q) = \cos(\psi_p)\sin(\psi_q) + \sin(\psi_p)\cos(\psi_q)$. Thus, we can write the (p, q)th element of the upper triangular matrix $\widetilde{\mathbf{R}}_g$ as

$$\widetilde{\mathbf{R}}_g(p,q) = \begin{cases} \Re\big(\mathbf{R}_g(p,q)\big) + \jmath \Im\big(\mathbf{R}_g(p,q)\big), & p > q \\ 1, & p = q. \end{cases} \tag{3.31}$$

By utilizing the information of $\widetilde{\mathbf{U}}$, the constrained optimization problem in Eq. (3.27) can be transformed into an unconstrained optimization problem that can be written as Eq. (3.28), where

$$\boldsymbol{\Theta} = \left[\boldsymbol{\psi}^T \; \widetilde{\boldsymbol{\psi}}^T \; \alpha\right]^T, \tag{3.32}$$

and

$$\begin{aligned} \boldsymbol{\psi}^T &= \left[\psi_{21} \; \psi_{31} \; \cdots \; \psi_{M1}\right]^T, \\ \widetilde{\boldsymbol{\psi}}^T &= \left[\psi_1 \; \psi_2 \; \cdots \; \psi_M\right]^T. \end{aligned}$$

The optimization is over $M(M-1)/2 + M$ elements ψ_{mn} and ψ_l. The advantage of this approach lies in the free selection of elements of $\boldsymbol{\Theta}$ without affecting the positive semi-definite property and diagonal elements of $\widetilde{\mathbf{R}}_g$. Noting that $\widetilde{\mathbf{U}}$ and $\widetilde{\mathbf{R}}_g$ are functions of $\boldsymbol{\Theta}$, we can alternately write the cost-function, in Eq. (3.28), as

$$J(\boldsymbol{\Theta}) = \frac{1}{K}\sum_{k=1}^{K}\Bigg[\frac{2}{\pi}\mathbf{a}^H(\theta_k)\sin^{-1}\Big(\Re(\mathbf{R}_g)\Big)\mathbf{a}(\theta_k) + \frac{2\jmath}{\pi}\mathbf{a}^H(\theta_k)\sin^{-1}\Big(\Im(\mathbf{R}_g)\Big)\mathbf{a}(\theta_k) - \alpha\phi(\theta_k)\Bigg]^2. \tag{3.33}$$

First, the partial differentiation of $J(\boldsymbol{\Theta})$ with respect to any element of $\boldsymbol{\psi}$, say ψ_{mn}, can be found as

$$\frac{\partial J(\boldsymbol{\Theta})}{\partial \psi_{mn}} = \Bigg[\frac{2}{K}\sum_{k=1}^{K}\Bigg\{\frac{2}{\pi}\mathbf{a}^H(\theta_k)\sin^{-1}\big(\Re(\mathbf{R}_g)\big)\mathbf{a}(\theta_k) + \frac{2\jmath}{\pi}\mathbf{a}^H(\theta_k)\sin^{-1}\big(\Im(\mathbf{R}_g)\big)\mathbf{a}(\theta_k) - \alpha\phi(\theta_k)\Bigg\}\Bigg]\Bigg[\frac{\partial}{\partial\psi_{mn}}\Bigg\{\frac{2}{\pi}\mathbf{a}^H(\theta_k)\sin^{-1}\big(\Re(\mathbf{R}_g)\big)\mathbf{a}(\theta_k) + \frac{2\jmath}{\pi}\mathbf{a}^H(\theta_k) \sin^{-1}\big(\Im(\mathbf{R}_g)\big)\mathbf{a}(\theta_k)\Bigg\}\Bigg]. \tag{3.34}$$

The matrix $\Re(\mathbf{R}_g)$ is real and symmetric, i.e., $\Re\big(\mathbf{R}_g(p,q)\big) = \Re\big(\mathbf{R}_g(q,p)\big)$, at the same time, $\Im(\mathbf{R}_g)$ has real entries but is skew-symmetric, i.e., $\Im\big(\mathbf{R}_g(p,q)\big) = -\Im\big(\mathbf{R}_g(q,p)\big)$. These observations enables us to write Eq. (3.34) in a simpler form

$$\frac{\partial J(\boldsymbol{\Theta})}{\partial \psi_{mn}} = \left[\frac{4}{K} \sum_{k=1}^{K} \left\{ \frac{2}{\pi} \mathbf{a}^H(\theta_k) \sin^{-1}\left(\Re(\mathbf{R}_g)\right)\mathbf{a}(\theta_k) + \frac{2\jmath}{\pi} \mathbf{a}^H(\theta_k) \sin^{-1}\left(\Im(\mathbf{R}_g)\right) \right.\right.$$
$$\left.\left. \mathbf{a}(\theta_k) - \alpha\phi(\theta_k) \right\}\right]\left[\frac{2}{\pi} \sum_{p=1}^{M-1} \sum_{q=p+1}^{M} \frac{\cos\left(\pi|p-q|\sin(\theta_k)\right)}{\sqrt{1-\Re\left(\mathbf{R}_g^2(p,q)\right)}} \frac{\partial \Re\left(\mathbf{R}_g(p,q)\right)}{\partial \psi_{mn}} \right]. \tag{3.35}$$

where $\partial \Re\left(\mathbf{R}_g(p,q)\right)/\partial \psi_{mn}$ is the partial derivative of the (p,q)th entry of $\Re(\mathbf{R}_g)$ with respect to ψ_{mn}. Note that $\Re(\mathbf{R}_g)$ contains only $(M-1)$ terms which depend on ψ_{mn}, thus, Eq. (3.35) further simplifies as

$$\frac{\partial J(\boldsymbol{\Theta})}{\partial \psi_{mn}} = \frac{8}{\pi K} \left[\sum_{k=1}^{K} \left\{ \frac{2}{\pi} \mathbf{a}^H(\theta_k) \sin^{-1}\left(\Re(\mathbf{R}_g)\right)\mathbf{a}(\theta_k) + \frac{2\jmath}{\pi} \mathbf{a}^H(\theta_k) \right.\right.$$
$$\left.\left. \sin^{-1}\left(\Im(\mathbf{R}_g)\right)\mathbf{a}(\theta_k) - \alpha\phi(\theta_k) \right\}\right]\left[\left\{ \sum_{p=1}^{m-1} \frac{\cos\left(\pi|p-m|\sin(\theta_k)\right)}{\sqrt{1-\Re\left(\mathbf{R}_g^2(p,m)\right)}} \frac{\partial \Re\left(\mathbf{R}_g(p,m)\right)}{\partial \psi_{mn}} \right.\right.$$
$$\left.\left. + \sum_{q=m+1}^{M} \frac{\cos\left(\pi|m-q|\sin(\theta_k)\right)}{\sqrt{1-\Re\left(\mathbf{R}_g^2(m,q)\right)}} \frac{\partial \Re\left(\mathbf{R}_g(m,q)\right)}{\partial \psi_{mn}} \right\}\right]. \tag{3.36}$$

Second, the partial differentiation of $J(\boldsymbol{\Theta})$ with respect to any element of $\widetilde{\psi}$, say ψ_l, can be found in the same manner as was found for ψ_{mn}, i.e.,

$$\frac{\partial J(\boldsymbol{\Theta})}{\partial \psi_l} = \frac{8}{\pi K} \left[\sum_{k=1}^{K} \left\{ \frac{2}{\pi} \mathbf{a}^H(\theta_k) \sin^{-1}\left(\Re(\mathbf{R}_g)\right)\mathbf{a}(\theta_k) + \frac{2\jmath}{\pi} \mathbf{a}^H(\theta_k) \sin^{-1}\left(\Im(\mathbf{R}_g)\right) \right.\right.$$
$$\left.\left. \mathbf{a}(\theta_k) - \alpha\phi(\theta_k) \right\}\right]\left[\sum_{p=1}^{M-1} \sum_{q=p+1}^{M} \frac{\cos\left(\pi|p-q|\sin(\theta_k)\right)}{\sqrt{1-\Re\left(\mathbf{R}_g^2(p,q)\right)}} \frac{\partial \Re\left(\mathbf{R}_g(p,q)\right)}{\partial \psi_l} \right]. \tag{3.37}$$

Finally, the partial differentiation of $J(\boldsymbol{\Theta})$ with respect to α is

$$\frac{\partial J(\boldsymbol{\Theta})}{\partial \alpha} = \frac{-2\phi(\theta_k)}{K} \left[\sum_{k=1}^{K} \left\{ \frac{2}{\pi} \mathbf{a}^H(\theta_k) \sin^{-1}\left(\Re(\mathbf{R}_g)\right)\mathbf{a}(\theta_k) + \frac{2\jmath}{\pi} \mathbf{a}^H(\theta_k) \right.\right.$$
$$\left.\left. \sin^{-1}\left(\Im(\mathbf{R}_g)\right)\mathbf{a}(\theta_k) - \alpha\phi(\theta_k) \right\}\right]. \tag{3.38}$$

3.4 Waveform Design for Spectrum Sharing

In the previous section, we designed finite alphabet constant-envelope QPSK waveforms by solving a beampattern matching optimization problem. In this section, we extend the beampattern matching optimization problem and introduce new constraints in order to tailor waveforms that don't cause interference to communication systems when MIMO radar and communication systems are sharing spectrum. We design spectrum sharing waveforms for two cases: the first case is for a stationary maritime MIMO radar and the second case is for moving maritime MIMO radar. The process of waveform design and its performance is discussed in the next sections.

3.4.1 Stationary Maritime MIMO Radar

Consider a naval ship docked at the harbor. The radar mounted on top of that ship is also stationary. The interference channels are also stationary due to non-movement of ship and BSs. In such a scenario, the CSI has little to no variations and thus it is feasible to include the constraint of NSP, Eq. (3.40), into the optimization problem. Thus, the new optimization problem is formulated as

$$\min_{\psi_{ij},\psi_l} \frac{1}{K}\sum_{k=1}^{K}\left[\frac{2}{\pi}\mathbf{a}^H(\theta_k)\mathbf{P}_i\left\{\sin^{-1}\left(\Re(\widetilde{\mathbf{U}})^H\Re(\widetilde{\mathbf{U}})+\Im(\widetilde{\mathbf{U}})^H\Im(\widetilde{\mathbf{U}})\right)\right.\right.$$
$$\left.\left.+\, \jmath\sin^{-1}\left(\Re(\widetilde{\mathbf{U}})^H\Im(\widetilde{\mathbf{U}})-\Im(\widetilde{\mathbf{U}})^H\Re(\widetilde{\mathbf{U}})\right)\right\}\mathbf{P}_i^H\mathbf{a}^H(\theta_k)-\alpha\phi(\theta_k)\right]^2. \quad (3.39)$$

A drawback of this approach is that it does not guarantee to generate constant-envelope radar waveform. However, the designed waveform is in the null space of the interference channel, thus, satisfying spectrum sharing constraints. The waveform generation process is shown using the block diagram of Fig. 3.1. Note that, $\mathcal{K}$ waveforms are designed, as we have $\mathcal{K}$ interference channels that are static. Using the projection matrix $\mathbf{P}_i$, the NSP projected waveform can be obtained as

$$\breve{\widetilde{\mathbf{Z}}}^{\text{opt}}_{\text{NSP}} = \widetilde{\mathbf{Z}}_i^{\text{opt}}\mathbf{P}_i^H. \quad (3.40)$$

The correlation matrix of the NSP waveform is given as

$$\breve{\widetilde{\mathbf{R}}}_i = \frac{1}{N}\left(\breve{\widetilde{\mathbf{Z}}}^{\text{opt}}_{\text{NSP}}\right)^H\breve{\widetilde{\mathbf{Z}}}^{\text{opt}}_{\text{NSP}}. \quad (3.41)$$

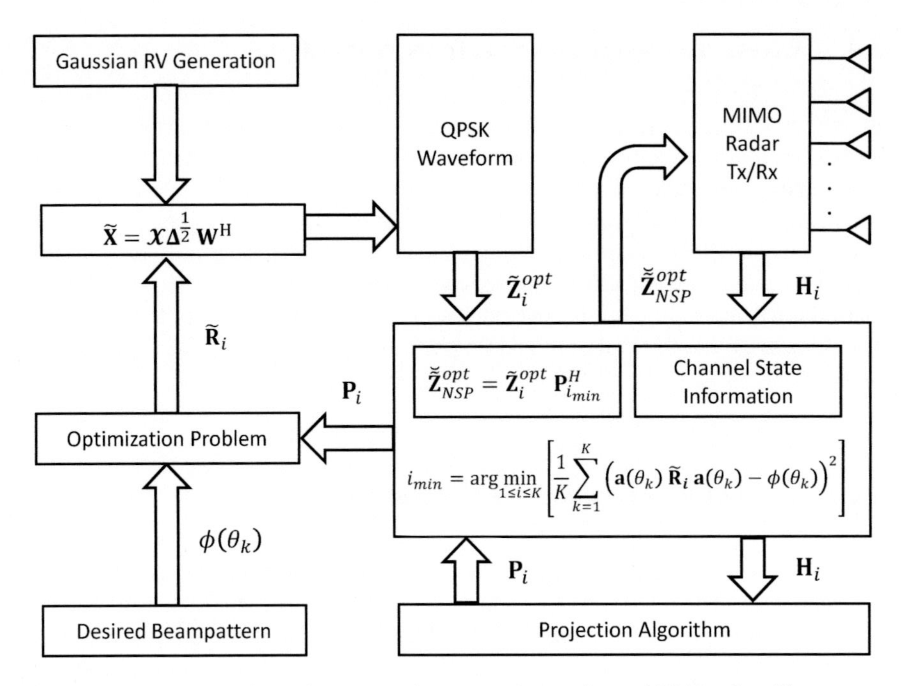

Fig. 3.1 Block diagram of waveform generation process for a stationary MIMO radar with spectrum sharing constraints

We propose to select the transmitted waveform with covariance matrix $\breve{\widetilde{\mathbf{R}}}_i$ is as close as possible to the desired covariance matrix, i.e.,

$$i_{\min} \triangleq \underset{1\le i\le \mathcal{K}}{\operatorname{argmin}} \left[\frac{1}{K} \sum_{k=1}^{K} \left(\mathbf{a}^H(\theta_k)\, \breve{\widetilde{\mathbf{R}}}_i\, \mathbf{a}(\theta_k) - \phi(\theta_k) \right)^2 \right] \tag{3.42}$$

$$\widetilde{\mathbf{R}}_{\mathrm{NSP}}^{\mathrm{opt}} \triangleq \breve{\widetilde{\mathbf{R}}}_{i_{\min}}. \tag{3.43}$$

Equivalently, we select $\mathbf{P}_i$ which projects maximum power at target locations. Thus, for stationary MIMO radar waveform with spectrum sharing constraints we propose Algorithm (4).

3.4.2 Moving Maritime MIMO Radar

Consider the case of a moving naval ship. The radar mounted on top of the ship is also moving, thus, the interference channels are varying due to the motion of ship. Due to

Algorithm 4 Stationary MIMO Radar Waveform Design Algorithm with Spectrum Sharing Constraints

loop
  for $i = 1 : \mathcal{K}$ **do**
    Get CSI of $\mathbf{H}_i$ through feedback from the ith BS.
    Send $\mathbf{H}_i$ to Algorithm (1) for the formation of projection matrix $\mathbf{P}_i$.
    Receive the ith projection matrix $\mathbf{P}_i$ from Algorithm (1).
    Design QPSK waveform $\widetilde{\mathbf{Z}}_i^{\text{opt}}$ using the optimization problem in Eq. (3.39).
    Project the QPSK waveform onto the null space of ith interference channel using $\breve{\widetilde{\mathbf{Z}}}_{\text{NSP}}^{\text{opt}} = \widetilde{\mathbf{Z}}_i^{\text{opt}}\mathbf{P}_i^H$.
  end for
  Find $i_{\min} = \underset{1\leq i\leq\mathcal{K}}{\text{argmin}}\left[\frac{1}{K}\sum_{k=1}^{K}\left(\mathbf{a}^H(\theta_k)\,\breve{\widetilde{\mathbf{R}}}_i\,\mathbf{a}(\theta_k) - \phi(\theta_k)\right)^2\right]$.
  Set $\widetilde{\mathbf{R}}_{\text{NSP}}^{\text{opt}} = \breve{\widetilde{\mathbf{R}}}_{i_{\min}}$ as the covariance matrix of the desired NSP QPSK waveforms to be transmitted.
end loop

time-varying ICSI, it is not feasible to include the NSP in the optimization problem. For this case, we first design finite alphabet constant-envelope QPSK waveforms, using the optimization problem in Eq. (3.28), and then use NSP to satisfy spectrum sharing constraints using transform

$$\breve{\widetilde{\mathbf{Z}}}_i = \widetilde{\mathbf{Z}}\mathbf{P}_i^H. \tag{3.44}$$

The waveform generation process is shown using the block diagram of Fig. 3.2. Note that only one waveform is designed using the optimization problem in Eq. (3.28) but $\mathcal{K}$ projection operations are performed via Eq. (3.44). The transmitted waveform is selected on the basis of minimum Frobenius norm with respect to the designed waveform using the optimization problem in Eq. (3.28), i.e.,

$$i_{\min} \triangleq \underset{1\leq i\leq\mathcal{K}}{\text{argmin}}\,||\widetilde{\mathbf{Z}}\mathbf{P}_i^H - \widetilde{\mathbf{Z}}||_F \tag{3.45}$$

$$\breve{\widetilde{\mathbf{Z}}}_{\text{NSP}} \triangleq \breve{\widetilde{\mathbf{Z}}}_{i_{\min}}. \tag{3.46}$$

The correlation matrix of this transmitted waveform is given as

$$\widetilde{\mathbf{R}}_{\text{NSP}} = \frac{1}{N}\breve{\widetilde{\mathbf{Z}}}_{\text{NSP}}^H\breve{\widetilde{\mathbf{Z}}}_{\text{NSP}}. \tag{3.47}$$

Thus, for moving MIMO radar waveform with spectrum sharing constraints we propose Algorithm (5).

Algorithm 5 Moving MIMO Radar Waveform Design Algorithm with Spectrum Sharing Constraints

Design FACE QPSK waveform $\widetilde{\mathbf{Z}}$ using the optimization problem in Eq. (3.28).
loop
  for $i = 1 : \mathcal{K}$ **do**
    Get CSI of $\mathbf{H}_i$ through feedback from the ith BS.
    Send $\mathbf{H}_i$ to Algorithm (1) for the formation of projection matrix $\mathbf{P}_i$.
    Receive the ith projection matrix $\mathbf{P}_i$ from Algorithm (1).
    Project the FACE QPSK waveform onto the null space of ith interference channel using $\breve{\widetilde{\mathbf{Z}}}_i = \widetilde{\mathbf{Z}}\mathbf{P}_i^H$.
  end for
  Find $i_{\min} = \underset{1 \leq i \leq \mathcal{K}}{\operatorname{argmin}} ||\widetilde{\mathbf{Z}}\mathbf{P}_i^H - \widetilde{\mathbf{Z}}||_F$.
  Set $\widetilde{\mathbf{R}}_{\text{NSP}}$ as the covariance matrix of the desired NSP QPSK waveforms to be transmitted.
end loop

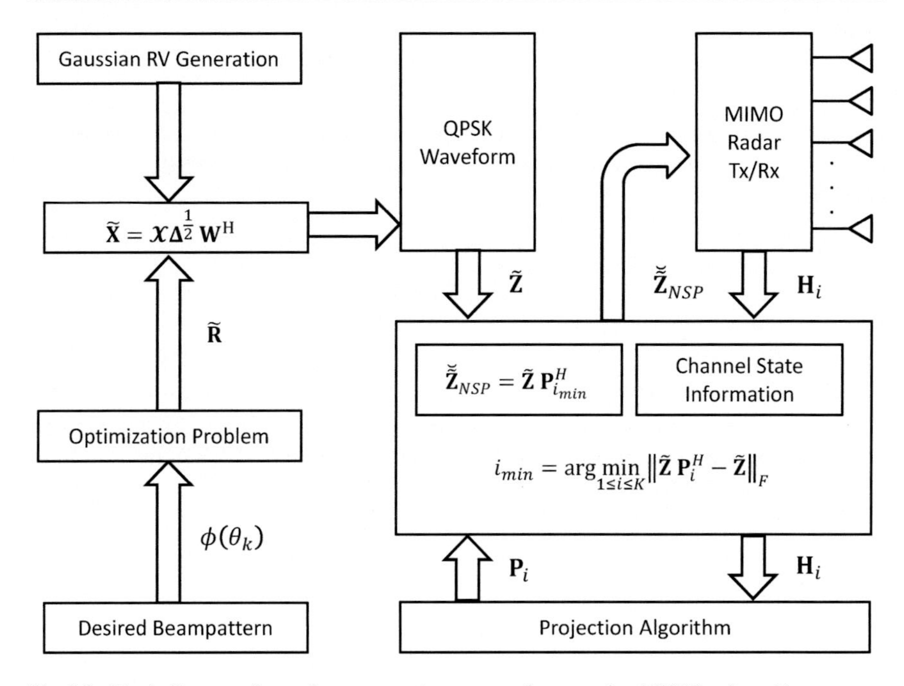

Fig. 3.2 Block diagram of waveform generation process for a moving MIMO radar with spectrum sharing constraints

3.4.3 *Performance Analysis*

There are different metrics to analyze performance of the designed waveform. These include analysis of the Woodward ambiguity function of waveform or looking at performance metrics such as angle of arrival or target detection capabilities of the designed waveform. In [4, 5] it is shown that uncorrelated orthogonal waveforms are optimal for target localization. For NSP waveforms, the performance of the projected/designed waveforms is comprehensively studied in [6–14]. The results show minimal degradation in radar performance when its waveform is subjected to null space projection. Although, Woodward ambiguity function has not been considered thus far to evaluate the performance of NSP BPSK/QPSK waveform but this can be done along the lines of [15–18].

3.5 Numerical Examples

In order to design QPSK waveforms with spectrum sharing constraints, we use a uniform linear array (ULA) of ten elements, i.e., $M = 10$, with an inter-element spacing of half-wavelength. Each antenna transmits waveform with unit power and $N = 100$ symbols. We average the resulting beampattern over 100 Monte-Carlo trials of QPSK waveforms. At each run of Monte Carlo simulation we generate a Rayleigh interference channel with dimensions $N_{\text{BS}} \times M$, calculate its null space, and solve the optimization problem for stationary and moving maritime MIMO radar.

3.5.1 Waveform for Stationary Radar

In this section, we design the transmit beampattern for a stationary MIMO radar. The desired beampattern has two main lobes from $-60°$ to $-40°$ and from $40°$ to $60°$. The QPSK transmit beampattern for stationary maritime MIMO radar is obtained by solving the optimization problem in Eq. (3.39). We give different examples to cover various scenarios involving different number of BSs and different configuration of MIMO antennas at the BSs. We also give one example to demonstrate the efficacy of Algorithms (1) and (4) in BS selection and its impact on the waveform design problem.

Example 1: Cellular System with five BSs and $\{3, 5, 7\}$ MIMO antennas and stationary MIMO radar
In this example, we design waveform for a stationary MIMO radar in the presence of a cellular system with five BSs. We look at three cases where we vary the number of BS antennas from $\{3, 5, 7\}$. In Fig. 3.3, we show the designed waveforms for all five BSs each equipped with 3 MIMO antennas. Note that, due to channel variations there is a large variation in the amount of power projected onto target locations for different BSs. But for certain BSs, the projected waveform is close to the desired waveform. In Fig. 3.4, we show the designed waveforms for all five BSs each equipped with 5 MIMO antennas. Similar to the previous case, due to channel variations there is a large variation in the amount of power projected onto target locations for different BSs. However, the power projected onto the target is less when compared with the previous case. We increase the number of antennas to 7 in Fig. 3.5, and notice that

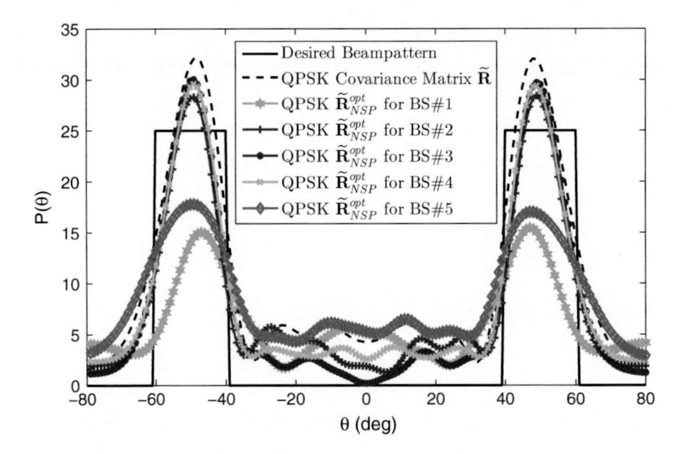

Fig. 3.3 QPSK waveform for stationary MIMO radar, sharing RF environment with five BSs each equipped with *three* antennas

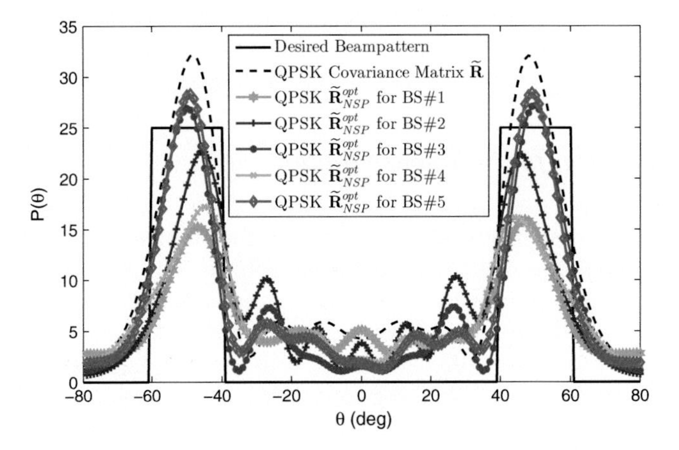

Fig. 3.4 QPSK waveform for stationary MIMO radar, sharing RF environment with five BSs each equipped with *five* antennas

the amount of power projected onto the targets is least as compared to previous two cases. This is because when $N_{\text{BS}} \ll M$ we have a larger null space to project radar waveform and this results in the projected waveform closer to the desired waveform. However, when $N_{\text{BS}} < M$, this is not the case.

Example 2: Performance of Algorithms (1) and (4) in BS selection for spectrum sharing with stationary MIMO radar

In Examples 1, we designed waveforms for different number of BSs with different antenna configurations. As we showed, for some BSs the designed waveform was close to the desired waveform but for other it wasn't and the projected waveform was closer to the desired waveform when $N_{\text{BS}} \ll M$ then when $N_{\text{BS}} < M$. In Fig. 3.6,

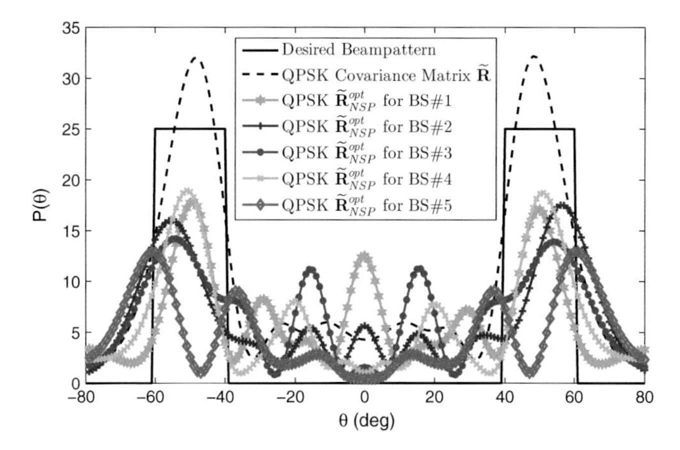

Fig. 3.5 QPSK waveform for stationary MIMO radar, sharing RF environment with five BSs each equipped with *seven* antennas

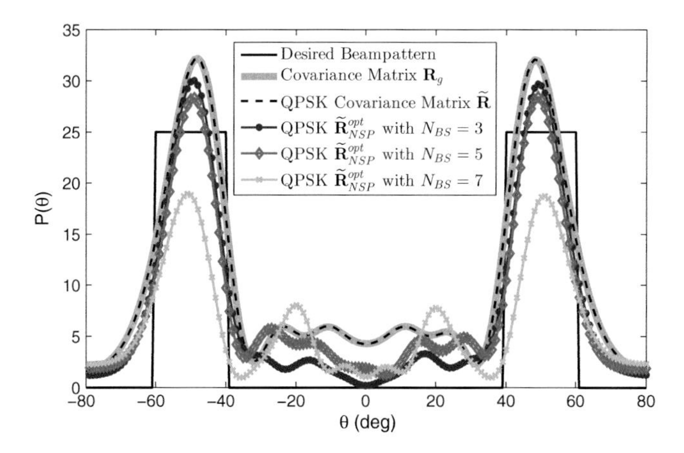

Fig. 3.6 Algorithm (4) is used to select the waveform which projects maximum power on the targets when $N_{\text{BS}} = \{3, 5, 7\}$ in the presence of five BSs

we use Algorithms (1) and (4) to select the waveform which projects maximum power on the targets or equivalently the projected waveform is closest to the desired waveform. We apply Algorithms (1) and (4) to the cases when $N_{\text{BS}} = \{3, 5, 7\}$ and select the waveform which projects maximum power on the targets. It can be seen that Algorithm (4) helps us to select waveform for stationary MIMO radar which results in best performance for radar in terms of projected waveform as close as possible to the desired waveform in addition of meeting spectrum sharing constraints.

3.5.2 *Waveform for Moving Radar*

In this section, we design transmit beampattern for a moving MIMO radar. The desired beampattern has two main lobes from -60° to -40° and from 40° to 60°. The QPSK transmit beampattern for moving maritime MIMO radar is obtained by solving the optimization problem in Eq. (3.33) and then projecting the resulting waveform onto the null space of $\mathbf{H}_i$ using the projection matrix in Eq. (3.44). We give different examples to cover various scenarios involving different number of BSs and different configuration of MIMO antennas at the BSs. We also give one example to demonstrate the efficacy of Algorithms (1) and (5) in BS selection and its impact on the waveform design problem.

Example 3: Cellular System with five BSs each with $\{3, 5, 7\}$ MIMO antennas and moving MIMO radar
In this example, we design waveform for a moving MIMO radar in the presence of a cellular system with five BSs. We look at three cases where we vary the number of BS antennas from $\{3, 5, 7\}$. In Fig. 3.7, we show the designed waveforms for all five BSs each equipped with 3 MIMO antennas. Note that, due to channel variations there is a large variation in the amount of power projected onto target locations for different BSs. When compared with Fig. 3.3, the power projected onto the target by NSP waveform is less due to the mobility of radar. In Fig. 3.8, we show the designed waveforms for all five BSs each equipped with 5 MIMO antennas. Similar to the previous case, due to channel variations there is a large variation in the amount of power projected onto target locations for different BSs. However, the power projected onto the target is less when compared with the previous case. We increase the number of antennas to 7 in Fig. 3.9, and notice that the amount of power projected onto the targets is least as compared to previous two cases. This is because when $N_{\text{BS}} \ll M$

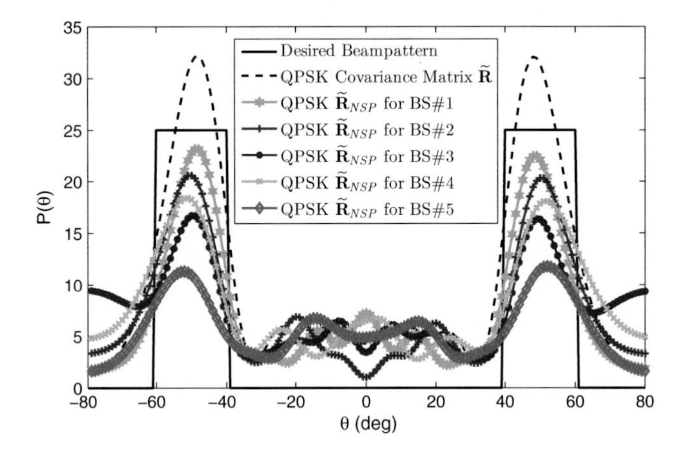

Fig. 3.7 QPSK waveform for moving MIMO radar, sharing RF environment with five BSs each equipped with *three* antennas

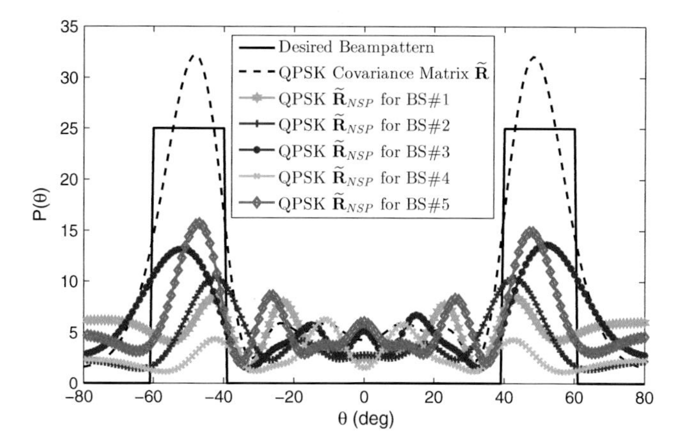

Fig. 3.8 QPSK waveform for moving MIMO radar, sharing RF environment with five BSs each equipped with *five* antennas

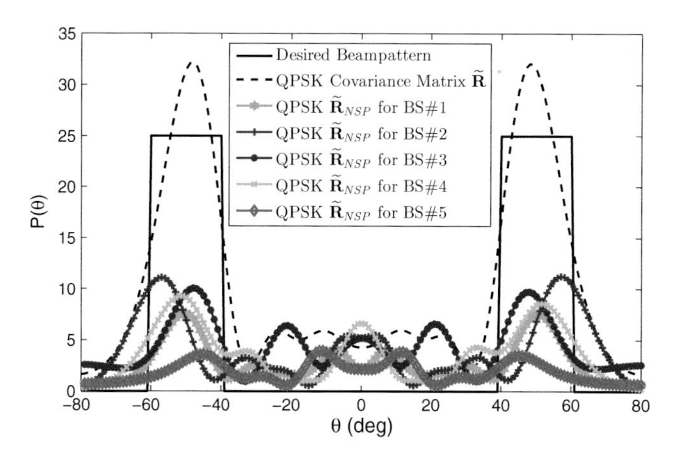

Fig. 3.9 QPSK waveform for moving MIMO radar, sharing RF environment with five BSs each equipped with *seven* antennas

we have a larger null space to project radar waveform and this results in the projected waveform closer to the desired waveform. However, when $N_{\text{BS}} < M$, this is not the case. Moreover, due to mobility of the radar, the amount of power projected for all three cases considered in this example are less than the similar example considered for stationary radar.

Example 4: Performance of Algorithms (1) and (5) in BS selection for spectrum sharing with moving MIMO radar

In Examples 3, we designed waveforms for different number of BSs with different antenna configurations. As we showed, for some BSs the designed waveform was close to the desired waveform but for other it wasn't and the projected waveform was

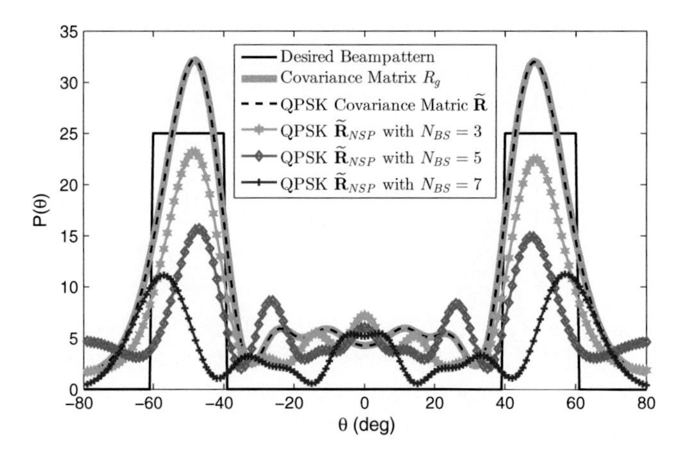

Fig. 3.10 Algorithm (4) is used to select the waveform which projects maximum power on the targets when $N_{\mathrm{BS}} = \{3, 5, 7\}$ in the presence of five BSs

closer to the desired waveform when $N_{\mathrm{BS}} \ll M$ then when $N_{\mathrm{BS}} < M$. In Fig. 3.10, we use Algorithms (1) and (5) to select the waveform which has the least Frobenius norm with respect to the designed waveform. We apply Algorithms (1) and (5) to the cases when $N_{\mathrm{BS}} = \{3, 5, 7\}$ and select the waveform which has minimum Frobenius norm. It can be seen that Algorithm (5) helps us to select waveform for stationary MIMO radar which results in best performance for radar in terms of projected waveform as close as possible to the desired waveform in addition of meeting spectrum sharing constraints.

3.6 Conclusion

Waveform design for MIMO radar is an active topic of research in the signal processing community. This work addressed the problem of designing MIMO radar waveforms with constant-envelope, which are very desirable from practical perspectives, and waveforms which allow radars to share spectrum with communication systems without causing interference, which are very desirable for spectrum congested RF environments.

In this chapter, we first showed that it is possible to realize finite alphabet constant-envelope quadrature-pulse shift keying (QPSK) MIMO radar waveforms. We proved that such the covariance matrix for QPSK waveforms is positive semi-definite and the constrained nonlinear optimization problem can be transformed into an un-constrained nonlinear optimization problem, to realize finite alphabet constant-envelope QPSK waveforms. This result is of importance for both communication and radar waveform designs where constant-envelope is highly desirable.

Second, we addressed the problem of radar waveform design for spectrally congested RF environments where radar and communication systems are sharing the same frequency band. We designed QPSK waveforms with spectrum sharing constraints. The QPSK waveform was shaped in a way that it is in the null space of communication system to avoid interference to communication system. We considered a multi-BS MIMO cellular system and proposed algorithms for the formation of projection matrices and selection of interference channels. We designed waveforms for stationary and moving MIMO radar systems. For stationary MIMO radar we presented an algorithm for waveform design by considering the spectrum sharing constraints. Our algorithm selected the waveform capable to project maximum power at the targets. For moving MIMO radar we presented another algorithm for waveform design by considering spectrum sharing constraints. Our algorithm selected the waveform with the minimum Frobenius norm with respect to the designed waveform. This metric helped to select the projected waveform closest to the designed waveform.

References

1. S. Ahmed, J. Thompson, Y. Petillot, B. Mulgrew, Unconstrained synthesis of covariance matrix for MIMO radar transmit beampattern. IEEE Trans. Signal Process. **59**, 3837–3849 (2011)
2. A. Hyvärinen, J. Karhunen, E. Oja, *Independent Component Analysis* (Wiley-Interscience, New York, 2001)
3. S. Sodagari, A. Abdel-Hadi, Constant envelope radar with coexisting capability with LTE communication systems. Under submission
4. I. Bekkerman, J. Tabrikian, Target detection and localization using mimo radars and sonars. IEEE Trans. Signal Process. **54**, 3873–3883 (2006)
5. Z. Qiu, J. Yang, H. Chen, X. Li, Z. Zhuang, Effects of transmitting correlated waveforms for co-located multi-input multi-output radar with target detection and localisation. IET Signal Process. **7**, 897–910 (2013)
6. S. Sodagari, A. Khawar, T.C. Clancy, R. McGwier, A projection based approach for radar and telecommunication systems coexistence, in *IEEE Global Communications Conference (GLOBECOM)* (2012)
7. A. Khawar, A. Abdel-Hadi, T.C. Clancy, Spectrum sharing between S-band radar and LTE cellular system: A spatial approach, in *IEEE DYSPAN* (2014)
8. A. Khawar, A. Abdel-Hadi, T.C. Clancy, R. McGwier, Beampattern analysis for MIMO radar and telecommunication system coexistence, in *IEEE ICNC* (2014)
9. A. Khawar, A. Abdel-Hadi, T.C. Clancy, On the impact of time-varying interference-channel on the spatial approach of spectrum sharing between S-band radar and communication system, in *IEEE MILCOM* (2014)
10. A. Khawar, A. Abdelhadi, T.C. Clancy, Performance analysis of coexisting radar and cellular system in LoS channel. Under submission
11. A. Khawar, A. Abdelhadi, T.C. Clancy, Channel modeling between MIMO seaborne radar and MIMO cellular system, in *CoRR* (2015). arXiv:abs/1504.04325
12. A. Khawar, A. Abdelhadi, T.C. Clancy, 3D channel modeling between seaborne radar and cellular system, in *CoRR* (2015). arXiv:abs/1504.04333
13. A. Khawar, A. Abdel-Hadi, T.C. Clancy, MIMO radar waveform design for coexistence with cellular systems, in *IEEE DYSPAN* (2014)

14. A. Khawar, A. Abdelhadi, T.C. Clancy, Target detection performance of spectrum sharing MIMO radars, in *Sensors Journal, IEEE*. **15**(9), 4928–4940, Sept 2015
15. G. Zhang, C. Wang, F. Chen, Automatic recognition of intra-pulse modulation type of radar signal based on ambiguity function, in *Recent Advances in Computer Science and Information Engineering*, vol. 128, Lecture Notes in Electrical Engineering, ed. by Z. Qian, L. Cao, W. Su, T. Wang, H. Yang (Springer, Berlin, 2012), pp. 659–664
16. P. Stinco, M. Greco, F. Gini, M. Rangaswamy, Ambiguity function and cramer-rao bounds for universal mobile telecommunications system-based passive coherent location systems. IET Radar Sonar Navig. **6**, 668–678 (2012)
17. J. Donohoe, F. Ingels, The ambiguity properties of FSK/PSK signals, in *IEEE Radar Conference* pp. 268–273, May 1990
18. N. Touati, C. Tatkeu, T. Chonavel, A. Rivenq, Phase coded costas signals for ambiguity function improvement and grating lobes suppression, in *78th IEEE Vehicular Technology Conference (VTC Fall)* pp. 1–5, Sept 2013

Appendix A
Solution of Waveform Design Optimization Problem

A.1 Preliminaries

This section presents some preliminary results used in the proofs throughout the paper. For proofs of the following theorems, please see the corresponding references.

Theorem A.1 *The matrix* $\mathbf{A} \in \mathbb{C}^{n \times n}$ *is positive semi-definite if and only if* $\Re(\mathbf{A})$ *is positive semi-definite [49].*

Theorem A.2 *A necessary and sufficient condition for* $\mathbf{A} \in \mathbb{C}^{n \times n}$ *to be positive definite is that the Hermitian part*

$$\mathbf{A}_H = \frac{1}{2}\left[\mathbf{A} + \mathbf{A}^H\right]$$

be positive definite [49].

Theorem A.3 *If* $\mathbf{A} \in \mathbb{C}^{n \times n}$ *and* $\mathbf{B} \in \mathbb{C}^{n \times n}$ *are positive semi-definite matrices then the matrix* $\mathbf{C} = \mathbf{A} + \mathbf{B}$ *is guaranteed to be positive semi-definite matrix [50].*

Theorem A.4 *If the matrix* $\mathbf{A} \in \mathbb{C}^{n \times n}$ *is positive semi-definite then the* p *times Schur product of* $\mathbf{A}$*, denoted by* $\mathbf{A}_{\circ}^{p}$*, will also be positive semi-definite [50].*

A.2 Generating CE QPSK Random Processes From Gaussian Random Variables

Assuming identically distributed Gaussian RV's $\widetilde{x}_p$, $\widetilde{y}_p$, $\widetilde{x}_q$ and $\widetilde{y}_q$ that are mapped onto QPSK RV's $\widetilde{z}_p$ and $\widetilde{z}_q$ using

A. Khawar et al., *MIMO Radar Waveform Design for Spectrum Sharing with Cellular Systems*, SpringerBriefs in Electrical and Computer Engineering,
DOI 10.1007/978-3-319-29725-5

$$\widetilde{z}_p = \frac{1}{\sqrt{2}}\left[\text{sign}\left(\frac{\widetilde{x}_p}{\sqrt{2}\sigma}\right) + \jmath\,\text{sign}\left(\frac{\widetilde{y}_p}{\sqrt{2}\sigma}\right)\right] \tag{A.1}$$

$$\widetilde{z}_q = \frac{1}{\sqrt{2}}\left[\text{sign}\left(\frac{\widetilde{x}_q}{\sqrt{2}\sigma}\right) + \jmath\,\text{sign}\left(\frac{\widetilde{y}_q}{\sqrt{2}\sigma}\right)\right] \tag{A.2}$$

where σ^2 is the variance of Gaussian RVs. The cross-correlation between QPSK and Gaussian RVs can be derived as

$$\mathbb{E}\{\widetilde{z}_p\widetilde{z}_q^*\} = \frac{1}{2}\mathbb{E}\Bigg[\left\{\text{sign}\left(\frac{\widetilde{x}_p}{\sqrt{2}\sigma}\right) + \jmath\,\text{sign}\left(\frac{\widetilde{y}_p}{\sqrt{2}\sigma}\right)\right\} \left\{\text{sign}\left(\frac{\widetilde{x}_q}{\sqrt{2}\sigma}\right) + \jmath\,\text{sign}\left(\frac{\widetilde{y}_q}{\sqrt{2}\sigma}\right)\right\}\Bigg]. \tag{A.3}$$

Using Eq. (3.12) we can write the above equation as

$$\mathbb{E}\{\widetilde{z}_p\widetilde{z}_q^*\} = \mathbb{E}\left\{\text{sign}\left(\frac{\widetilde{x}_p}{\sqrt{2}\sigma}\right)\text{sign}\left(\frac{\widetilde{x}_q}{\sqrt{2}\sigma}\right)\right\} + \jmath\,\mathbb{E}\left\{\text{sign}\left(\frac{\widetilde{y}_p}{\sqrt{2}\sigma}\right)\text{sign}\left(\frac{\widetilde{x}_q}{\sqrt{2}\sigma}\right)\right\}. \tag{A.4}$$

The cross-correlation relationship between Gaussian and QPSK RVs can be derived by first considering

$$\mathbb{E}\left\{\text{sign}\left(\frac{\widetilde{x}_p}{\sqrt{2}\sigma}\right)\text{sign}\left(\frac{\widetilde{x}_q}{\sqrt{2}\sigma}\right)\right\} = \int_{-\infty}^{\infty}\int_{-\infty}^{\infty}\left[\text{sign}\left(\frac{\widetilde{x}_p}{\sqrt{2}\sigma}\right)\text{sign}\left(\frac{\widetilde{x}_q}{\sqrt{2}\sigma}\right) p(\widetilde{x}_p, \widetilde{x}_q, \rho_{\widetilde{x}_p\widetilde{x}_q})\right] d\widetilde{x}_p\, d\widetilde{x}_q \tag{A.5}$$

where $p(\widetilde{x}_p, \widetilde{x}_q, \rho_{\widetilde{x}_p\widetilde{x}_q})$ is the joint probability density function of $\widetilde{x}_p$ and $\widetilde{x}_q$, and $\rho_{\widetilde{x}_p\widetilde{x}_q} = \frac{\mathbb{E}\{\widetilde{x}_p\widetilde{x}_q^*\}}{\sigma^2}$ is the cross-correlation coefficient of $\widetilde{x}_p$ and $\widetilde{x}_q$. Using Hermite polynomials [51], the above double integral can be transformed as in [8]. Thus,

$$\mathbb{E}\left\{\text{sign}\left(\frac{\widetilde{x}_p}{\sqrt{2}\sigma}\right)\text{sign}\left(\frac{\widetilde{x}_q}{\sqrt{2}\sigma}\right)\right\} = \sum_{n=0}^{\infty}\frac{\rho^n_{\widetilde{x}_p\widetilde{x}_q}}{2\pi\sigma^2 2^n n!}\int_{-\infty}^{\infty}\text{sign}\left(\frac{\widetilde{x}_p}{\sqrt{2}\sigma}\right)e^{\widetilde{x}_p^2/2\sigma^2} H_n\left(\frac{\widetilde{x}_p}{\sqrt{2}\sigma}\right) d\widetilde{x}_p \int_{-\infty}^{\infty}\text{sign}\left(\frac{\widetilde{x}_q}{\sqrt{2}\sigma}\right)e^{\widetilde{x}_q^2/2\sigma^2} H_n\left(\frac{\widetilde{x}_q}{\sqrt{2}\sigma}\right) d\widetilde{x}_q \tag{A.6}$$

where

$$H_n(\widetilde{x}_m) = (-1)^n e^{\frac{\widetilde{x}_m^2}{2}} \frac{d^n}{d\widetilde{x}_m^n} e^{\frac{-\widetilde{x}_m^2}{2}} \tag{A.7}$$

is the Hermite polynomial. By substituting $\hat{x}_p = \frac{\widetilde{x}_p}{\sqrt{2}\sigma}$ and $\hat{x}_q = \frac{\widetilde{x}_q}{\sqrt{2}\sigma}$, and splitting the limits of integration into two parts, Eq. (A.6) can be simplified as

$$\mathbb{E}\Big\{\operatorname{sign}(\hat{x}_p)\operatorname{sign}(\hat{x}_q)\Big\} = \sum_{n=0}^{\infty} \frac{\rho^n_{\hat{x}_p\hat{x}_q}}{\pi 2^n n!}\left(\int_0^{\infty} e^{\hat{x}_p^2}\Big[H_n(\hat{x}_p) - H_n(-\hat{x}_p)\Big]\,d\hat{x}_p\right)^2. \quad (A.8)$$

Using $H_n(-\hat{x}_p) = (-1)^n H_n(\hat{x}_p)$ [52], Eq. (A.8) can be written as

$$\mathbb{E}\Big\{\operatorname{sign}(\hat{x}_p)\operatorname{sign}(\hat{x}_q)\Big\} = \sum_{n=0}^{\infty} \frac{\rho^n_{\hat{x}_p\hat{x}_q}}{\pi 2^n n!}\left(\int_0^{\infty} e^{\hat{x}_p^2} H_n(\hat{x}_p)\big(1 - (-1)^n\big)\,d\hat{x}_p\right)^2. \quad (A.9)$$

The above equation is non-zero for odd n only, therefore, we can rewrite it as

$$\mathbb{E}\Big\{\operatorname{sign}(\hat{x}_p)\operatorname{sign}(\hat{x}_q)\Big\} = \sum_{n=0}^{\infty} \frac{\rho^{2n+1}_{\hat{x}_p\hat{x}_q}}{\pi 2^{2n}(2n+1)!}\left(\int_0^{\infty} e^{\hat{x}_p^2} H_{2n+1}(\hat{x}_p)\,d\hat{x}_p\right)^2. \quad (A.10)$$

Then using $\int_0^{\infty} e^{\hat{x}_p^2} H_{2n+1}(\hat{x}_p)\,d\hat{x}_p = (-1)^n \frac{(2n)!}{n!}$ from [52], we can write Eq. (A.10) as

$$\begin{aligned}\mathbb{E}\Big\{\operatorname{sign}\Big(\frac{\widetilde{x}_p}{\sqrt{2}\sigma}\Big)\operatorname{sign}\Big(\frac{\widetilde{x}_q}{\sqrt{2}\sigma}\Big)\Big\} &= \sum_{n=0}^{\infty} \frac{\rho^{2n+1}_{\widetilde{x}_p\widetilde{x}_q}}{\pi 2^{2n}(2n+1)!}\left((-1)^n\frac{2n!}{n!}\right)^2 \\ &= \frac{2}{\pi}\left[\rho_{\widetilde{x}_p\widetilde{x}_q} + \frac{\rho^3_{\widetilde{x}_p\widetilde{x}_q}}{2\cdot 3} + \frac{1\cdot 3\rho^5_{\widetilde{x}_p\widetilde{x}_q}}{2\cdot 4\cdot 5} + \frac{1\cdot 3\cdot 5\rho^7_{\widetilde{x}_p\widetilde{x}_q}}{2\cdot 4\cdot 6\cdot 7} + \cdots\right] \\ &= \frac{2}{\pi}\sin^{-1}\Big(\mathbb{E}\{\widetilde{x}_p\widetilde{x}_q\}\Big) \qquad (A.11)\end{aligned}$$

In Eq. (A.5), we expanded the first part of Eq. (A.4). Now, similarly expanding the second part of Eq. (A.4), i.e.,

$$\begin{aligned}\mathbb{E}\Big\{\operatorname{sign}\Big(\frac{\widetilde{y}_p}{\sqrt{2}\sigma}\Big)\operatorname{sign}\Big(\frac{\widetilde{x}_q}{\sqrt{2}\sigma}\Big)\Big\} = \int_{-\infty}^{\infty}\int_{-\infty}^{\infty}\Big[&\operatorname{sign}\Big(\frac{\widetilde{y}_p}{\sqrt{2}\sigma}\Big)\operatorname{sign}\Big(\frac{\widetilde{x}_q}{\sqrt{2}\sigma}\Big) \\ &p(\widetilde{y}_p, \widetilde{x}_q, \rho_{\widetilde{y}_p\widetilde{x}_q})\Big]\,d\widetilde{y}_p\,d\widetilde{x}_q \qquad (A.12)\end{aligned}$$

where $p(\widetilde{y}_p, \widetilde{x}_q, \rho_{\widetilde{y}_p\widetilde{x}_q})$ is the joint probability density function of $\widetilde{y}_p$ and $\widetilde{x}_q$, and $\rho_{\widetilde{y}_p\widetilde{x}_q} = \frac{\mathbb{E}\{\widetilde{y}_p\widetilde{x}_q^*\}}{\sigma^2}$ is the cross-correlation coefficient of $\widetilde{y}_p$ and $\widetilde{x}_q$. Using Hermite polynomials, Eq. (A.7), we can write Eq. (A.12) as

$$\mathbb{E}\left\{\text{sign}\left(\frac{\widetilde{y}_p}{\sqrt{2}\sigma}\right)\text{sign}\left(\frac{\widetilde{x}_q}{\sqrt{2}\sigma}\right)\right\} = \sum_{n=0}^{\infty}\frac{\rho^n_{\widetilde{y}_p\widetilde{x}_q}}{2\pi\sigma^2 2^n n!}\int_{-\infty}^{\infty}\text{sign}\left(\frac{\widetilde{y}_p}{\sqrt{2}\sigma}\right)e^{\widetilde{y}_p^2/2\sigma^2}$$

$$H_n\left(\frac{\widetilde{y}_p}{\sqrt{2}\sigma}\right)d\widetilde{y}_p\int_{-\infty}^{\infty}\text{sign}\left(\frac{\widetilde{x}_q}{\sqrt{2}\sigma}\right)e^{\widetilde{x}_q^2/2\sigma^2}H_n\left(\frac{\widetilde{x}_q}{\sqrt{2}\sigma}\right)d\widetilde{x}_q. \tag{A.13}$$

By substituting $\hat{y}_p = \frac{\widetilde{y}_p}{\sqrt{2}\sigma}$ and $\hat{x}_q = \frac{\widetilde{x}_q}{\sqrt{2}\sigma}$, and splitting the limits of integration into two parts, Eq. (A.13) can be simplified as

$$\mathbb{E}\left\{\text{sign}(\hat{y}_p)\text{sign}(\hat{x}_q)\right\} = \sum_{n=0}^{\infty}\frac{\rho^n_{\hat{y}_p\hat{x}_q}}{\pi 2^n n!}\left(\int_0^{\infty}e^{\hat{y}_p^2}\left[H_n(\hat{y}_p) - H_n(-\hat{y}_p)\right]d\hat{y}_p\right)^2. \tag{A.14}$$

Using $H_n(-\hat{y}_p) = (-1)^n H_n(\hat{y}_p)$, above equation can be written as

$$\mathbb{E}\left\{\text{sign}(\hat{y}_p)\text{sign}(\hat{x}_q)\right\} = \sum_{n=0}^{\infty}\frac{\rho^n_{\hat{y}_p\hat{x}_q}}{\pi 2^n n!}\left(\int_0^{\infty}e^{\hat{y}_p^2}H_n(\hat{y}_p)\big(1-(-1)^n\big)\,d\hat{y}_p\right)^2. \tag{A.15}$$

The above equation is non-zero for odd n only, therefore, we can rewrite it as

$$\mathbb{E}\left\{\text{sign}(\hat{y}_p)\text{sign}(\hat{x}_q)\right\} = \sum_{n=0}^{\infty}\frac{\rho^{2n+1}_{\hat{y}_p\hat{x}_q}}{\pi 2^{2n}(2n+1)!}\left(\int_0^{\infty}e^{\hat{y}_p^2}H_{2n+1}(\hat{y}_p)\,d\hat{y}_p\right)^2. \tag{A.16}$$

Then using $\int_0^{\infty}e^{\hat{y}_p^2}H_{2n+1}(\hat{y}_p)\,d\hat{y}_p = (-1)^n\frac{(2n)!}{n!}$, we can write Eq. (A.16) as

$$\begin{aligned}\mathbb{E}\left\{\text{sign}\left(\frac{\widetilde{y}_p}{\sqrt{2}\sigma}\right)\text{sign}\left(\frac{\widetilde{x}_q}{\sqrt{2}\sigma}\right)\right\} &= \sum_{n=0}^{\infty}\frac{\rho^{2n+1}_{\widetilde{y}_p\widetilde{x}_q}}{\pi 2^{2n}(2n+1)!}\left((-1)^n\frac{2n!}{n!}\right)^2\\ &= \frac{2}{\pi}\left[\rho_{\widetilde{y}_p\widetilde{x}_q} + \frac{\rho^3_{\widetilde{y}_p\widetilde{x}_q}}{2\cdot 3} + \frac{1\cdot 3\rho^5_{\widetilde{y}_p\widetilde{x}_q}}{2\cdot 4\cdot 5} + \frac{1\cdot 3\cdot 5\rho^7_{\widetilde{y}_p\widetilde{x}_q}}{2\cdot 4\cdot 6\cdot 7} + \cdots\right]\\ &= \frac{2}{\pi}\sin^{-1}\left(\mathbb{E}\{\widetilde{y}_p\widetilde{x}_q\}\right).\end{aligned} \tag{A.17}$$

Combining Eqs. (A.11) and (A.17), gives us the cross-correlation of Eq. (A.4) as

$$\mathbb{E}\{\widetilde{z}_p\widetilde{z}_q\} = \frac{2}{\pi}\left[\sin^{-1}\left(\mathbb{E}\{\widetilde{x}_p\widetilde{x}_q\}\right) + \jmath\,\sin^{-1}\left(\mathbb{E}\{\widetilde{y}_p\widetilde{x}_q\}\right)\right]. \tag{A.18}$$

A.3 Proofs

Proof (Proof of Lemma 3.1) To prove Lemma 3.1, we note that the real part of $\widetilde{\mathbf{R}}_g$ is $\mathbf{R}_g$ which is positive semi-definite by definition, thus, by Theorem A.1, the complex covariance matrix $\widetilde{\mathbf{R}}_g$ is also positive semi-definite.

Proof (Proof of Lemma 3.2) To prove Lemma 3.2, we can individually expand the sum, $\sin^{-1}\left(\Re(\widetilde{\mathbf{R}}_g)\right) + \jmath \sin^{-1}\left(\Im(\widetilde{\mathbf{R}}_g)\right)$, using Taylor series, *i.e.*, first expanding $\sin^{-1}\left(\Re(\widetilde{\mathbf{R}}_g)\right)$

$$\sin^{-1}\left(\Re(\mathbf{R}_g)\right) = \Re(\mathbf{R}_g) + \frac{1}{2\cdot 3}\Re(\mathbf{R}_g)^3_\circ + \frac{1\cdot 3}{2\cdot 4\cdot 5}\Re(\mathbf{R}_g)^5_\circ + \frac{1\cdot 3\cdot 5}{2\cdot 4\cdot 6\cdot 7}\Re(\mathbf{R}_g)^7_\circ + \cdots \quad \text{(A.19)}$$

Then using Theorem A.3, each term or matrix, on the right hand side, is positive semi-definite, since, $\Re(\mathbf{R}_g)$ is positive semi-definite by definition. Moreover, $\sin^{-1}\left(\Re(\mathbf{R}_g)\right)$ is also positive semi-definite since its a sum of positive semi-definite matrices, this follows from Theorem A.1.

Similarly, expanding $\jmath \sin^{-1}\left(\Im(\mathbf{R}_g)\right)$ as

$$\jmath \sin^{-1}\left(\Im(\mathbf{R}_g)\right) = \jmath[\Im(\mathbf{R}_g) + \frac{1}{2\cdot 3}\Im(\mathbf{R}_g)^3_\circ + \frac{1\cdot 3}{2\cdot 4\cdot 5}\Im(\mathbf{R}_g)^5_\circ + \frac{1\cdot 3\cdot 5}{2\cdot 4\cdot 6\cdot 7}\Im(\mathbf{R}_g)^7_\circ + \cdots] \quad \text{(A.20)}$$

Now, $\widetilde{\mathbf{R}}$ is positive semi-definite since real part of it is positive semidefinite, from Eq. (A.19) and Theorem A.4.

References

1. D.S. Bernstein, *Matrix Mathematics: Theory, Facts, and Formulas*, second edn. (Princeton University Press, 2009)
2. R.A. Horn, C.R. Johnson, *Matrix Analysis*, (Cambridge University Press, Cambridge, U.K., 1985)
3. J. Brown, Jr., On the expansion of the bivariate gaussian probability density using results of nonlinear theory (corresp.). IEEE Trans. Inf. Theor. **14**, 158–159, Sept 1968
4. S. Ahmed, J.S. Thompson, Y.R. Petillot, B. Mulgrew, Finite alphabet constant-envelope waveform design for MIMO radar. IEEE Trans. Signal Process. **59**(11), 5326–5337 (2011)
5. A. De Maio, S. De Nicola, A. Farina, S. Iommelli, Adaptive detection of a signal with angle uncertainty. IET Radar, Sonar Navig. **4**, 537–547, Aug 2010

Appendix B
MATLAB Code for Waveform Design

B.1 BPSK Waveform Design

In this section, we provide MATLAB code to generate BPSK radar waveforms for stationary and moving MIMO radar platforms subject to spectrum sharing constraints.

B.1.1 Stationary Maritime MIMO Radar

The following script designs BPSK radar waveform for stationary platforms that are subject to spectrum sharing constraints.

```
clear all
close all
clc
global phi P_v_breve

N_s = 100;          % Number of samples
M_T = 10;           % MIMO Radar Tx Antennas
M_R = M_T;;         % MIMO Radar Rx Antennas
N_R = 10;           % Communication System Rx Antennas
alpha = 1;          % Scaling Factor
sigma2_w = 1;       % Noise Variacne
Sig_threshold = 2;
l = 1;
K = 181;            % Number of theta points on graph [-90,90]
k1 = -55;           % Starting angle with nonzero desired value
k2 = -45;           % Ending angle with nonzero desired value
k3 = 45;            % Starting angle with nonzero desired value
                    % (phase 2)
k4 = 55;            % Ending angle with nonzero desired value
                    % (phase 2)
x0 = zeros(1,45);   % initial value for optimization
```


A. Khawar et al., *MIMO Radar Waveform Design for Spectrum Sharing with Cellular Systems*, SpringerBriefs in Electrical and Computer Engineering,
DOI 10.1007/978-3-319-29725-5

```
x00 = zeros(1,10);
x1 = -pi/1 * ones(1,45);
x2 = pi/1 * ones(1,45);
x3 = -pi/1 * ones(1,10);
x4 = pi/1 * ones(1,10);
H_iterations = 100;  % Number of H realizations;
N_iterations = 100;  % Number of Gaussian Generator realizations;
% Desired Beamform
phi = zeros(1,K);    % Desired o/p beamform 1:181 = = -90:90
for i4 = k1 + 91:k2 + 91
phi(i4) = 25;
end
for i4 = k3 + 91:k4 + 91
phi(i4) = 25;
end
phi;

%%%%% Optimization for no null BPSK case %%%%%%
theta_noNull = fminsearchbnd(@SQP, [x0, 1],[x1, 1], [x2, 1])% SQP
U = Spherical_form(theta_noNull);
R_noNull = (2/pi) * (asin(ctranspose(U) * U));      % CE matrix R
[S_noNull V_noNull D_noNull] = svd(R_noNull);

%%%% For different Sigma Threshold %%%%
for kk2 = 1:3
Sig_threshold = 2 * kk2;
Sig_threshold_2(kk2) = Sig_threshold;
%%%% 100 Monte Carlo trials for Gaussian Generator  %%%%
for kk1 = 1: N_iterations
% Gaussian Matrix M_T * N
N = random('normal',0,sigma2_w,N_s,M_T);
% Gaussian Waveform generation
X = N * (V_noNull)^(1/2) * ctranspose(S_noNull);
% Corresponding BPSK waveform
Z = sign(real(X));
% BPSK matrix R_BPSK
R_BPSK = (1/N_s) * ctranspose(Z) * Z;
% Drawing the beampattern for no null projection case
for i6 =1:181
theta_2 = (pi/180) * (i6 - 91);
theta_3(i6) = (i6 - 91);
a_TT = [1; exp(i * pi *1* sin(theta_2)); ...
exp(i * pi *2* sin(theta_2)); ...
exp(i * pi *3* sin(theta_2)); exp(i * pi *4* sin(theta_2));...
exp(i * pi *5* sin(theta_2)); exp(i * pi *6* sin(theta_2));...
exp(i * pi *7* sin(theta_2)); exp(i * pi *8* sin(theta_2));...
exp(i * pi *9* sin(theta_2))];
P(i6,kk1)= abs(ctranspose(a_TT) * R_noNull * (a_TT));
% covariance matrix
P_BPSK_noNull(i6,kk1)= abs(ctranspose(a_TT) * ...
R_BPSK * (a_TT)); % BPSK
end
%%% Monte Carlo Simulation for Channel  %%%
for(j = 1 : H_iterations)
%%%% Null Space Projection  %%%%
% Singular Value Decomposition of Channel State H
H = random('rayleigh',2,N_R,M_T);
 % parameter is (4-pi/4)*variance
[S V D] = svd(H);
```

```
% Null Space Projectoion of H to P_v for singular value less than
% Sigma_threshold
for (k = 1 : M_T)
if (V(k,k)<Sig_threshold)
V_breve (:,l) = D(:,k);
l = l + 1;
end
% Condition to aviod having no projection matrix
if (l == 1)
V_breve (:,l) = D(:,M_T);
end
end
V_breve;
%j
% Projection matrix into null space of H
P_v_breve = V_breve * inv(ctranspose(V_breve)* V_breve) *...
ctranspose(V_breve);
%%%% END (Null Space Projection)%%%%
% BPSK + null projection
R_BPSK_Null = P_v_breve *  R_BPSK * ctranspose(P_v_breve);
% Drawing the beampattern for null projection case
for i6 =1:181
theta_2 = (pi/180) * (i6 - 91);
theta_3(i6) = (i6 - 91);
a_TT = [1; exp(i * pi *1* sin(theta_2)); ...
exp(i * pi *2* sin(theta_2)); exp(i * pi *3* sin(theta_2));...
exp(i * pi *4* sin(theta_2)); exp(i * pi *5* sin(theta_2));...
exp(i * pi *6* sin(theta_2)); exp(i * pi *7* sin(theta_2));...
exp(i * pi *8* sin(theta_2)); exp(i * pi *9* sin(theta_2))];
P_BPSK_Null(i6,kk2, kk1,j)= abs(ctranspose(a_TT) * ...
R_BPSK_Null * (a_TT));
end
end % channel Monte Carlo Simulation

P_BPSK_Null_av = mean(P_BPSK_Null,4); % avarage over channel
end  % Gaussian Random Generator Monte Carlo Simulation
P_av = mean(P,2);   % before BPSK waveform (soln of opt Rg)
% after BPSK waveform
% average over noise (after BPSK waveform and NSP)
P_BPSK_noNull_av = mean(P_BPSK_noNull,2);
P_BPSK_Null_av_av = mean(P_BPSK_Null_av,3);
Error_BPSK(:,kk2) = (1/181) * ...
(P_BPSK_Null_av_av(:,kk2) - phi').^2
end % the sigma threshold
Error_BPSK_av = mean(Error_BPSK)
subplot(1,2,1)
plot(theta_3, phi,'b', theta_3, P_BPSK_Null_av_av)
xlabel('\theta (deg)')
ylabel('P(\theta)')
legend('Desired Beam', 'threshold=2','threshold=4','threshold=6')
subplot(1,2,2)
plot(Sig_threshold_2, 10 * log(Error_BPSK_av))
xlabel('threshold')
ylabel('10 log(MSE)')
```

B.1.2 Moving Maritime MIMO Radar

This section provides MATLAB code for BPSK waveform when radar platform is subject to motion.

```
clear all
close all
clc
global phi P_v_breve

N_s = 100;          % Number of samples
M_T = 10;           % MIMO Radar Tx Antennas
M_R = M_T;          % MIMO Radar Rx Antennas
N_R = 10;           % Communication System Rx Antennas
alpha = 1;          % Scaling Factor
sigma2_w = 1;       % Noise Variacne
Sig_threshold = 2;
l = 1;
K = 181;            % Number of theta points on graph [-90,90]
k1 = -55;           % Starting angle with nonzero desired value
k2 = -45;           % Ending angle with nonzero desired value
k3 = 45;            % Starting angle with nonzero desired value
                    % (phase 2)
k4 = 55;            % Ending angle with nonzero desired value
                    % (phase 2)
x0 = zeros(1,45);% initial value for optimization
x00 = zeros(1,10);
x1 = -pi/1 * ones(1,45);
x2 = pi/1 * ones(1,45);
x3 = -pi/1 * ones(1,10);
x4 = pi/1 * ones(1,10);
H_iterations = 100;% Number of H realizations;
N_iterations = 100;% Number of Gaussian Generator realizations;
% Desired Beamform
phi = zeros(1,K);  % Desired o/p beamform 1:181 = = -90:90
for i4 = k1 + 91:k2 + 91
phi(i4) = 25;
end
for i4 = k3 + 91:k4 + 91
phi(i4) = 25;
end
phi;
%%%%% For different Sigma Threshold %%%%%
for kk2 = 1:3
Sig_threshold = 2 * kk2;
Sig_threshold_2(kk2) = Sig_threshold;
%%%%% Monte Carlo Simulation for Channel  %%%%%
for(j = 1 : H_iterations)
l = 1;
%%%% Null Space Projection  %%%%
% Singular Value Decomposition of Channel State H
H = random('rayleigh',2,N_R,M_T);
% parameter is (4-pi/4)*variance
[S V D] = svd(H); % Null Space Projectoion of H to P_v
% for singular value less than Sigma_threshold
for (k = 1 : M_T)
if (V(k,k)<Sig_threshold)
```

```
V_breve (:,l) = D(:,k);
l = l + 1;
end
% Condition to aviod having no projection matrix
if (l == 1)
V_breve (:,l) = D(:,M_T);
end
end
V_breve;
% Projection matrix into null space of H
P_v_breve = V_breve * inv(ctranspose(V_breve)* V_breve) ...
* ctranspose(V_breve);
%%% END (Null Space Projection)%%%
%%% Optimization for NSP in optimization case %%%
theta_noNull = fminsearchbnd(@SQP_case2, ...
[x0, 1],[x1, 1], [x2, 1]);% SQP
U = Spherical_form(theta_noNull);
R_noNull = (2/pi) * (asin(ctranspose(U) * U));     % CE matrix R
[S_noNull V_noNull D_noNull] = svd(R_noNull);
%%%% 100 Monte Carlo trials for Gaussian Generator %%%%
for kk1 = 1: N_iterations
% Gaussian Matrix M_T * N
N = random('normal',0,sigma2_w,N_s,M_T);
% Gaussian Waveform generation
X = N * (V_noNull)^(1/2) * ctranspose(S_noNull);
% Corresponding BPSK waveform
Z = sign(real(X));
% BPSK matrix R_BPSK
R_BPSK(:,:,kk1) = (1/N_s) * ctranspose(Z) * Z;
R_BPSK_Null(:,:,kk1) = P_v_breve *  R_BPSK(:,:,kk1) ...
 * ctranspose(P_v_breve); % after BPSK waveform
end  % Gaussian Random Generator Monte Carlo Simulation
R_BPSK_Null_av(:,:,j) = mean(R_BPSK_Null,3);
% averaging over noise
end % channel Monte Carlo Simulation
% avarage over channel
R_BPSK_Null_av_av(:,:,kk2) = mean(R_BPSK_Null_av,3);
% Drawing the beampattern for null projection case
for i6 =1:181
theta_2 = (pi/180) * (i6 - 91);
theta_3(i6) = (i6 - 91);
a_TT =
[1; exp(i * pi *1* sin(theta_2)); ...
exp(i * pi *2* sin(theta_2)); ...
exp(i * pi *3* sin(theta_2)); exp(i * pi *4* sin(theta_2)); ...
exp(i * pi *5* sin(theta_2)); exp(i * pi *6* sin(theta_2)); ...
exp(i * pi *7* sin(theta_2)); exp(i * pi *8* sin(theta_2)); ...
exp(i * pi *9* sin(theta_2))];
P_BPSK_Null_av_av(i6,kk2)= abs(ctranspose(a_TT) * ...
R_BPSK_Null_av_av(:,:,kk2) * (a_TT));
end
Error_BPSK(:,kk2) = (1/181)*(P_BPSK_Null_av_av(:,kk2) - phi').^2;
end % the sigma threshold

Error_BPSK_av = mean(Error_BPSK)
subplot(1,2,1)
plot(theta_3, phi,'b', theta_3, P_BPSK_Null_av_av)
xlabel('\theta (deg)')
ylabel('P(\theta)')
```

```
legend('Desired Beam', 'threshold=2','threshold=4','threshold=6')
subplot(1,2,2)
plot(Sig_threshold_2, 10 * log(Error_BPSK_av))
xlabel(' threshold')
ylabel('10 log(MSE)')
```

B.2 QPSK Waveform Design

In this section, we provide MATLAB code to generate QPSK radar waveforms for stationary and moving MIMO radar platforms subject to spectrum sharing constraints.

B.2.1 Stationary Maritime MIMO Radar

In this section, we provide MATLAB code to generate radar waveform for stationary radar platforms.

```
clear all
close all
clc
global phi P_v_breve

N_s = 10000;    % Number of samples
M_T = 10;       % MIMO Radar Tx Antennas
M_R = M_T;      % MIMO Radar Rx Antennas
N_R = 3;        % Communication System Rx Antennas
alpha = 1;      % Scaling Factor
sigma2_w = 0.1;% Noise Variacne
Sig_threshold = 2;
l = 1;
% Optimization I.C.
x0 = zeros(1,45);  % initial value for optimization
x00 = zeros(1,10);
x1 = -pi/1 * ones(1,45);
x2 = pi/1 * ones(1,45);
x3 = -pi/1 * ones(1,10);
x4 = pi/1 * ones(1,10);
H_iterations = 1; % Number of H realizations;
N_iterations = 1; % Number of Gaussian Generator realizations;
K = 181;          % Number of theta points on graph [-90,90]
% Desired Beamform 1 parameters
k1 = -60;         % Starting angle with nonzero desired value
k2 = -40;         % Ending angle with nonzero desired value
k3 = 40;          % Starting angle with nonzero desired value
                  % (phase 2)
k4 = 60;          % Ending angle with nonzero desired value
                  % (phase 2)
for i4 = k1 + 91:k2 + 91
phi(i4) = 25;
end
```

```
for i4 = k3 + 91:k4 + 91
phi(i4) = 25;
end
%%%% Optimization for no null QPSK case %%%%
theta_noNull = fminsearchbnd(@SQP_complex2, [x0, 1, x00], ...
[x1, 1, x3], [x2, 1, x4])% SQP
U_re = Spherical_form_real(theta_noNull);
% Speherical form
U_im = Spherical_form_imag(theta_noNull);
% Speherical form
U = complex(U_re, U_im);
Rg = ctranspose(U) * U;
Rg_real = real(Rg);
Rg_imag = imag(Rg)
R_S = [Rg_real, Rg_imag; -Rg_imag, Rg_real]
R_real = real((2/pi) * (asin(Rg_real)));
R_imag = real((2/pi) * (asin(Rg_imag)))
R_complex = complex(R_real, R_imag);
R_noNull = R_complex%(2/pi) * ...
(asin(ctranspose(U) * U)); % CE matrix R
CE matrix R
[S_noNull V_noNull D_noNull] = svd(R_S);
%%%%% For different Sigma Threshold %%%%
for kk2 = 1:1 % sigma_th = 2, 4, 6, 8, 10

%%%%% 100 Monte Carlo trials for Gaussian Generator %%%%
for k1 = 1: N_iterations
% Gaussian Matrix M_T * N
N_S = random('normal',0,sigma2_w,N_s,2*M_T);
% Gaussian Waveform generation
S = N_S * (V_noNull)^(1/2) * ctranspose(S_noNull);
% Corresponding QPSK waveform
Z = sqrt(1/2) * complex(sign(S(:,1:10)),sign(S(:,11:20)));
R_QPSK = (1/N_s) * ctranspose(Z) * Z;
% Drawing the beampattern for no null projection case
for i6 =1:181
theta_2 = (pi/180) * (i6 - 91);
theta_3(i6) = (i6 - 91);
a_TT =
[1; exp(i * pi *1* sin(theta_2)); ...
exp(i * pi *2* sin(theta_2)); ...
exp(i * pi *3* sin(theta_2)); exp(i * pi *4* sin(theta_2)); ...
exp(i * pi *5* sin(theta_2)); exp(i * pi *6* sin(theta_2)); ...
exp(i * pi *7* sin(theta_2)); exp(i * pi *8* sin(theta_2)); ...
exp(i * pi *9* sin(theta_2))];
P(i6,k1)= abs(ctranspose(a_TT) * R_noNull * (a_TT));
P_QPSK_noNull(i6,k1)= abs(ctranspose(a_TT) * R_QPSK * (a_TT));
end
P_av = mean(P,2);
P_QPSK_noNull_av = mean(P_QPSK_noNull,2);
%%% Monte Carlo Simulation for Channel %%%
for(j = 1 : H_iterations)
l = 1;
QPSK_noise_iterations = k1
Channel_iterations = j
%%%% Null Space Projection %%%%%
% Singular Value Decomposition of Channel State H
H = randn(N_R, M_T) + j * randn(N_R, M_T);
% Null Space Projectoion of H to P_v for singular
```

```
% value less than
P_v_breve = null(H) * ctranspose(null(H));
%%%% Optimization for null QPSK case  %%%%
theta_noNull = fminsearchbnd(@SQP_complex2_case2, ...
[x0, 1, x00],[x1, 1, x3], [x2, 1, x4])% SQP
U_re = Spherical_form_real(theta_noNull);
% Speherical form
U_im = Spherical_form_imag(theta_noNull);
% Speherical form
U = complex(U_re, U_im);
Rg = ctranspose(U) * U;
Rg_real = real(Rg);
Rg_imag = imag(Rg)
R_real = real((2/pi)  *  (asin(Rg_real)));
R_imag = real((2/pi)  *  (asin(imag(Rg))));
R_complex = P_v_breve * complex(R_real, R_imag) ...
 * ctranspose(P_v_breve);
R_Null =  R_complex
%%%% END (Null Space Projection) %%%%

% Drawing the beampattern for null projection case
for i6 =1:181
theta_2 = (pi/180) * (i6 - 91);
theta_3(i6) = (i6 - 91);
a_TT =
[1; exp(i * pi *1* sin(theta_2)); ...
 exp(i * pi *2* sin(theta_2)); ...
 exp(i * pi *3* sin(theta_2)); exp(i * pi *4* sin(theta_2)); ...
 exp(i * pi *5* sin(theta_2)); exp(i * pi *6* sin(theta_2)); ...
 exp(i * pi *7* sin(theta_2)); exp(i * pi *8* sin(theta_2)); ...
 exp(i * pi *9* sin(theta_2))];
P_QPSK_Null(i6,kk2, k1,j)=
abs(ctranspose(a_TT) * R_Null * (a_TT));
end % beampattern drawing
end  % Channel Monte Carlo Simulation
P_QPSK_Null_av = mean(P_QPSK_Null,4);% averageing over channel
end  % Gaussian Random Generator Monte Carlo Simulation
P_QPSK_Null_av_av = mean(P_QPSK_Null_av,3);
% averaging over noise
Error_QPSK(:,kk2) = (1/181) * ...
(P_QPSK_Null_av_av(:,kk2) - phi').^2   % MSE calculation
end % the sigma threshold
Error_QPSK_av = mean(Error_QPSK)
 % averaging over theta to get the total error
plot(theta_3, phi,'b', theta_3, P_av,'r', theta_3, ...
P_QPSK_noNull_av, 'g', theta_3, P_QPSK_Null_av_av, 'm')
legend('DESIRED','R','R_{QPSK}','PROJECTED QPSK')
```

B.2.2 Moving Maritime MIMO Radar

In this section, we provide MATLAB code to design MIMO radar waveform for moving radar platforms subject to spectrum sharing constraints.

```
clear all
close all
clc
global phi P_v_breve

N_s = 10000;  % Number of samples of QPSK
M_T = 10;     % MIMO Radar Tx Antennas (In this simulation it
% has to be 10 because of the spherical coordinates)
M_R = M_T;    % MIMO Radar Rx Antennas
N_R = 3;      % Communication System Rx Antennas
%(could be changed)
alpha = 1;    % Scaling Factor
sigma2_w = 0.1; % Noise Variacne
%Sig_threshold = 2;
l = 1;
% Optimization I.C.
x0 = zeros(1,45); % initial value for optimization
x00 = zeros(1,10);
x1 = -pi/1 * ones(1,45);
x2 = pi/1 * ones(1,45);
x3 = -pi/1 * ones(1,10);
x4 = pi/1 * ones(1,10);
H_iterations = 1;% Number of H realizations;
N_iterations = 1;% Number of Gaussian Generator realizations;
K = 181; % Number of theta points on graph [-90,90]
% Desired Beamform 1 parameters
k1 = -60;        % Starting angle with nonzero desired value
k2 = -40;        % Ending angle with nonzero desired value
k3 = 40;         % Starting angle with nonzero desired value
%(phase 2)
k4 = 60;         % Ending angle with nonzero desired value
%(phase 2)
% Desired Beamform 2 parameters
phi = zeros(1,K);% Desired o/p beamform 1:181 = = -90:90
% Desired Beamform 1
for i4 = k1 + 91:k2 + 91
phi(i4) = 25;
end
for i4 = k3 + 91:k4 + 91
phi(i4) = 25;
end

%%%%% Optimization for no null QPSK case  %%%%%
theta_noNull = fminsearchbnd(@SQP_complex2, [x0, 1, x00], ...
[x1, 1, x3], [x2, 1, x4])% SQP
U_re = Spherical_form_real(theta_noNull);
% Speherical form
U_im = Spherical_form_imag(theta_noNull);
% Speherical form
U = complex(U_re, U_im);
Rg = ctranspose(U) * U;
Rg_real = real(Rg);
Rg_imag = imag(Rg)
R_S = [Rg_real, Rg_imag; -Rg_imag, Rg_real]
R_real = real((2/pi)  *  (asin(Rg_real)));
R_imag = real((2/pi)  *  (asin(Rg_imag)))
R_complex = complex(R_real, R_imag);
R_noNull =  R_complex%(2/pi) * ...
(asin(ctranspose(U) * U));      % CE matrix R
```

```
[S_noNull V_noNull D_noNull] = svd(R_S);
%%%% For different number of samples %%%%
for kk2 = 1:1:1
%%%% 100 Monte Carlo trials for Gaussian Generator  %%%%
for k1 = 1: N_iterations

% Gaussian Waveform generation
N_S = random('normal',0,sigma2_w,N_s,2*M_T);
S = N_S * (V_noNull)^(1/2) * ctranspose(S_noNull);
% Corresponding QPSK waveform
Z = sqrt(1/2) * complex(sign(S(:,1:M_T)), ...
sign(S(:,M_T + 1:2*M_T)));
% QPSK matrix R_QPSK
R_QPSK = (1/N_s) * ctranspose(Z) * Z  ;
% Drawing the beampattern for no null projection case
for i6 =1:181
theta_2 = (pi/180) * (i6 - 91);
theta_3(i6) = (i6 - 91);
a_TT =
[1; exp(i * pi *1* sin(theta_2)); ...
 exp(i * pi *2* sin(theta_2)); ...
exp(i * pi *3* sin(theta_2)); exp(i * pi *4* sin(theta_2)); ...
exp(i * pi *5* sin(theta_2)); exp(i * pi *6* sin(theta_2)); ...
exp(i * pi *7* sin(theta_2)); exp(i * pi *8* sin(theta_2)); ...
exp(i * pi *9* sin(theta_2))];
P(i6,k1)= abs(ctranspose(a_TT) * R_noNull * (a_TT));
P_QPSK_noNull(i6,k1)= abs(ctranspose(a_TT) * ...
R_QPSK * (a_TT));
end

P_av = mean(P,2);
P_QPSK_noNull_av = mean(P_QPSK_noNull,2)
P_QPSK_noNull_av(:,kk2) = mean(P_QPSK_noNull,2)
Error_QPSKnoNull(:,kk2) = (1/181)* ...
(P_QPSK_noNull_av(:,kk2) - phi').^2
%%%% Monte Carlo Simulation for Channel  %%%%
for(j = 1 : H_iterations)
%%%% Null Space Projection  %%%%%
% Singular Value Decomposition of Channel State H
H = randn(N_R, M_T) + j * randn(N_R, M_T);
%%%%%% new null space projection %%%%%%
null(H);
P_v_breve = null(H) * ctranspose(null(H));
%%% END (Null Space Projection) %%%
R_QPSK_Null = P_v_breve *  R_QPSK  * ctranspose(P_v_breve);
% Drawing the beampattern for null projection case
for i6 =1:181
theta_2 = (pi/180) * (i6 - 91);
theta_3(i6) = (i6 - 91);
a_TT =
[1; exp(i * pi *1* sin(theta_2));
... exp(i * pi *2* sin(theta_2)); ...
exp(i * pi *3* sin(theta_2)); exp(i * pi *4* sin(theta_2)); ...
exp(i * pi *5* sin(theta_2)); exp(i * pi *6* sin(theta_2)); ...
exp(i * pi *7* sin(theta_2)); exp(i * pi *8* sin(theta_2)); ...
exp(i * pi *9* sin(theta_2))];
P_QPSK_Null(i6, kk2, k1, j)= abs(ctranspose(a_TT) ...
* R_QPSK_Null * (a_TT));
end
```

```
P_QPSK_Null_av = mean(P_QPSK_Null,4);
end % channel Monte Carlo Simulation
P_QPSK_Null_av_av = mean(P_QPSK_Null_av,3);
Error_QPSK(:,kk2) = (1/181)*(P_QPSK_Null_av_av(:,kk2) - phi').^2
end % Gaussian Random Generator Monte Carlo Simulation
end % the sigma threshold
Error_QPSK_av = mean(Error_QPSK)
Error_QPSKnoNull_av = mean(Error_QPSKnoNull)
plot(theta_3, phi,'b', theta_3, P_av,'r', theta_3, ...
P_QPSK_noNull_av, 'g', theta_3, P_QPSK_Null_av_av)
```

Printed in the United States
By Bookmasters